CLOCKWORK

Also by Richard Balzer

CHINA DAY BY DAY

(with Eileen Hsü-Balzer and Francis L. K. Hsü)

NEXT DOOR, DOWN THE ROAD, AROUND THE CORNER

STREET TIME

LEASE
CLOSE DOORS
HOT
DRIVE CAREFULLY

RICHARD BALZER

CLOCKWORK

Life in and Outside an American Factory

Doubleday & Company, Inc.
Garden City, New York
1976

ISBN 0-385-11036-7
Library of Congress Catalog Card Number 75-21209

Printed in the United States of America
BOOK DESIGN BY BENTE HAMANN
First Edition

TO CHARLES AND SELMA BALZER, MY PARENTS.

EACH TIME I THINK I'M A SELF-MADE PERSON,

I REALIZE I'M NOT.

Acknowledgments

There are some books that grow out of an author's imagination, and acknowledgments seem superfluous. There are other books, like this one, where one feels a deep obligation to thank many people without whose assistance such an undertaking would have been impossible.

For a while I worried that I would never get inside a factory. With the help of Wolf Berthold I met three people at the Western Electric plant in North Andover, Massachusetts, Harry Youngman, Jack Driscoll, and John Connors, all of whom helped in getting the project approved. The final decision was Dave Hilder's, general manager of the Merrimack Valley Works. I want to thank him for taking the risk of allowing this study to procede.

Although the union—Communications Workers of America—was not initially involved in the decision about my employment, the officers of Local 1365 were very helpful. I'd particularly like to thank Frank Talarico, the local's vice-president.

My baptism into factory work was out on the shop floor. Some of the people whom I met and learned from speak in their own voices in this book. Many more, however, go nameless. To all the people in the shops who not only taught me about work, and shared their lives with me, but who made me feel comfortable and accepted, I want to say thank you.

I am indebted to the Institute of Current World Affairs for the generous grant they gave me which supported this study. They also required me to write monthly newsletters about my experiences, and although often while writing them I hated the discipline, the newsletters have served as a foundation for the chapters in this book.

I want to thank Sam Vaughan, President of the Publishing Division of Doubleday, who took an early interest in my writing about Western Electric and encouraged me to expand it into a book.

Encouragement and support came from many different people. In the company's public relations department Louise Richer, Jim Chasse, and Don McDermott were especially generous with both their time and energy. Of all the people inside the plant who looked at my newsletters and gave me suggestions as to how to improve them, I'd particularly like to thank Barbara Vaughn and "the professor."

Once the material was in manuscript form, several people, including Patricia and Brenden Sexton, Peter Almond, Joan Liem, Taylor Stoehr, Duncan Kennedy, Diane Matthews, my editor at Doubleday, and Joe Lyons, helped move it from a rough draft to its final form.

Joe Lyons played a vital role in this study. He replaced John Connors as head of public relations. He was responsible for "reviewing" all my writing and giving me the "company view." He fullfilled the role, but over time a more important relationship developed. We became friends, and I'd like to thank him for his efforts to help make this as good a book as possible.

This book, as long as it is, has been distilled from notes, from interviews, and from seemingly countless drafts. I wish to thank Debi Smith, who typed that material and waded through the often barely intelligible tapes. Although I printed most of the pictures, I did call on Sven Martson to make some prints.

This book would have been impossible without the help of my wife, Eileen. She has been involved in every stage of its development. She has helped edit the manuscript and select photographs. She has acted as a cheering section when depression was at hand. She continues to be the most important person in my life.

Contents

CLOCKWORK

INTRODUCTION

Several years ago a fifty-year-old friend described a boyhood memory. On social occasions, my friend recalled, his father would stand around with his hands stuffed deep in his pockets, rarely removing them even to shake hands. My friend came to realize that his father, a factory worker, was ashamed of the stains which came from working with dyes.

Twenty-eight million men and women currently work in America's factories. Their lives, both at work and at home, remain hidden—like my friend's father's hands—from large segments of the American public.

I certainly come from that large segment. None of my childhood friends had parents who worked in factories. I never worked in a factory. It wasn't until serving as a VISTA volunteer and years later as a legal aid lawyer that I met and came to know people who worked in factories.

My first glimpse of what factory life was like came when I visited about a dozen factories while preparing a book about America. I realized it was a world I knew practically nothing about. In the late fall of 1972 I received a two-year grant to study the impact of social and economic change on "lower-middle-class Americans" from the Institute of Current World Affairs. One way to begin the research was to get a job working in a factory, so in the winter of 1973 I began looking for a factory job. The only thing I didn't want to do was to make up a phony name and personality. If I lied about my background and purpose on the job application, I'd also have to lie to the people I worked with. Even if successful, and I wasn't sure I would be, at the end I would have had to tell people that they had been deceived, and that was not the way I wanted to get to know them.

I knew that trying to get a company to go along with a plan which would permit me not only to work but later photograph in the plant wasn't going to be easy. For three months with the help of union officials, businessmen, and acquaintances, I looked for a job. The answer was always no. Sometimes the no was brusque—don't waste our time. Sometimes the no was more thoughtfully worded—we like your project, we're sorry we can't help you.

I was beginning to doubt that I would ever find a job my way, when some Western Electric officials at the Merrimack Valley Works in North Andover, Massachusetts, expressed some interest in my proposal.

Although my proposal met with an initial positive reaction, it took three months for a decision to be made. There were many lunches and meetings, a trip to Boston to talk to regional officials, and one to national headquarters in New York. There was an internal debate on the "wisdom" of such a project. Some local officials felt there was nothing to hide, and that they might learn something from my experience; others questioned my objectivity and feared unfavorable publicity if the study were biased.

I asked about including a union representative in the deliberations. A company representative informed me that the company would make a determination and should they decide to go along with the project they would clear it with the union before offering me a job.

Finally, when "all the bases had been touched," and everything had been considered, Western Electric took the chance. I was offered a job as a 32 grade (lowest level) bench hand. It was agreed that while I worked I would work like anyone else, and be paid like anyone else. I wouldn't take pictures until I finished working, and everyone in the group I worked with would be told from the beginning who I was and what I was going to do.

The plant's general manager also drew up an agreement containing the following provisions:

1) an employee's privacy and individual wishes will be of foremost concern at all times.
2) the company reserves the right to review and reject certain photographs, primarily those of a proprietary nature.
3) the company has the right to review material prior to publication and suggest corrections, resulting from misunderstandings, and also if deemed necessary the right to withhold the use of the company name.
4) the project may be terminated by either party at any time.

On May 29, 1973, I began work. I worked for four months on the first shift, which ran from 6:30 in the morning to 3:00 in the afternoon, in a relatively new miscellaneous equipment shop. I spent the fifth, and my last, month on the second shift, which ran from 3:00 P.M. to 11:30 P.M., in one of the thin film rooms.

Approximately ten thousand people are employed at the Merrimack Valley Works. Most of them reside in the greater Lawrence and Haverhill areas. There are about 6700 blue collar workers, 95 per cent are white, 58 per cent are women. Their average age is forty-seven, and their average length of service is twelve years. It is not uncommon to find more than one family member or more than one generation of a family working in the plant.

In the beginning I knew there was much more to work than just the tasks people performed. Aside from that I thought I had no other preconceptions about the work or the people I'd be working with. I immediately came to realize that I was bringing a whole bagful of preconceptions and misconceptions with me. For example, on my first day at work I noticed several women wearing pant suits or well-tailored dresses. Large numbers had their hair stylishly coiffed. None of this fit my unconscious stereotype of factory attire.

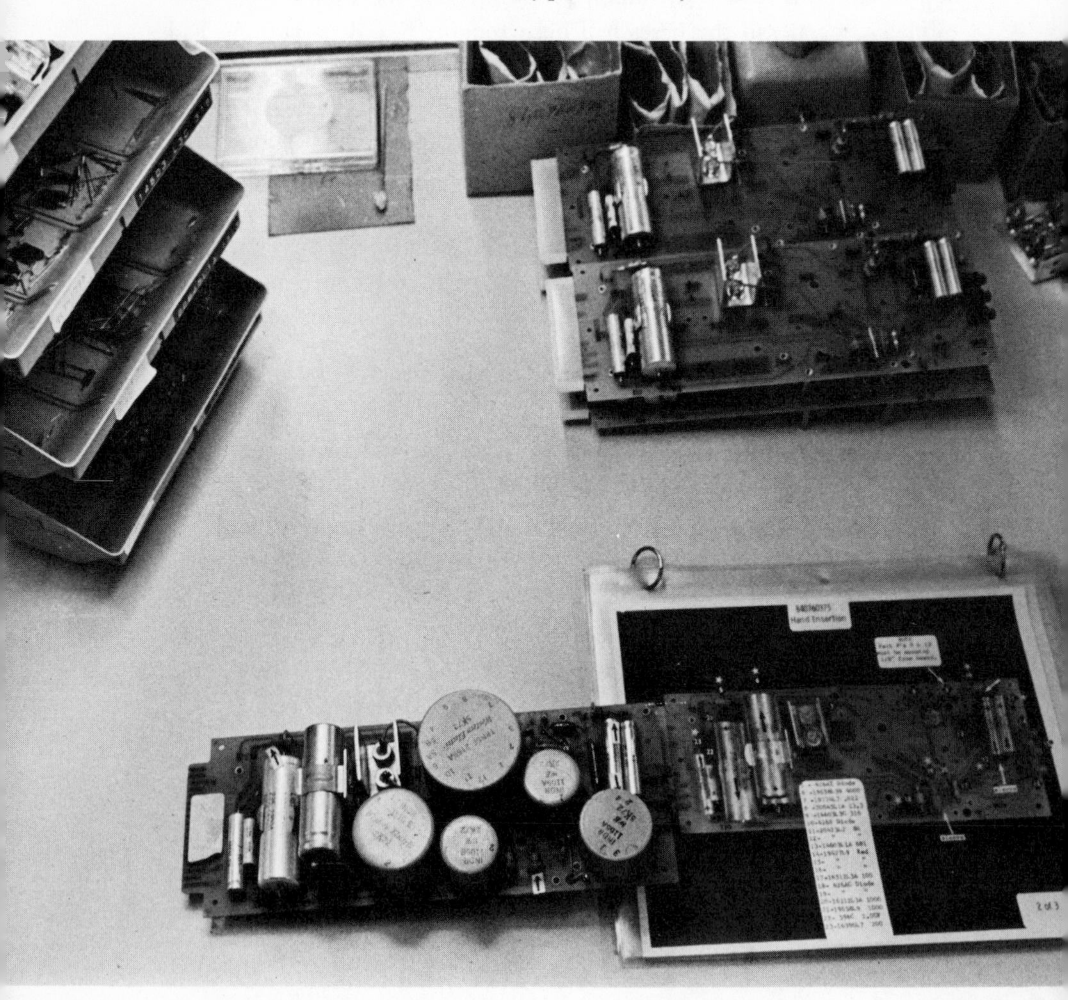

It didn't take long to discover that I would not quickly become a great worker. I handled my jobs much more clumsily than I had anticipated. Soon I realized that I had assumed, unconsciously but nevertheless condescendingly, that as soon as I had been taught a job I would be able to do it and do it well. I struggled for five months to get to the point where I could master the smallest of skills.

I never stopped uncovering my hidden preconceptions. The day before I stopped working on the first shift a group of women asked me if I thought I could, like them, hold a job like this for ten or fifteen years. I thought about it, and then said I hadn't minded the work, but that over such a long time I'd probably get angry and do something that would get me fired. One of the women looked at me and said, "That's a lot of bull. I watched you, and you'd do just like the rest of us. If you had to be here ten or fifteen years you'd figure out a way to do it."

She was probably right. I had assumed that I couldn't take the monotony, the sameness of the jobs, the authoritarianism. Yet experience taught me that most people can and do adjust to factory life, and if I had to I would too. I learned a great deal about myself working in the plant. This book reflects that education as much as it does other people's lives and reactions.

A substantial portion of this book is devoted to looking at the environment in which people work. Work is more than just tasks. It includes coffee breaks, lunch, coming to work early, staying late, talking, joking, teasing, breaking rules, and figuring out ways to beat the system. A person can have bad things to say about a company or a job, and still look forward to coming to work early. I have attempted to present what I have come to know about the reality of the workday of the people I worked with.

Most of my experiences come from the T2-M1, 2 miscellaneous group I worked with on the day shift. Our group produced two systems: T2, a digital carrier system, and M1, 2, a digital multiplexer. The T2 carrier system transmits 96 telephone conversations simultaneously over routes up to 500 miles long. The M1, 2 multiplexer is the equipment which combines conversations into one big stream for transmission. Our product was sold to the Bell Operating companies for intercity phone traffic. I make no claims that this one group is typical of all groups in this plant, or in the Western Electric system. I have used this experience as the focus of my writing about the company.

Despite the assurances the company gave me about not taking any punitive action against those who had negative things to say, many people were still afraid to say things about the company which would be credited to them. Most of the comments and quotes I have used in these chapters are not drawn from taped conversations but from work-day chatter, conversations in the johns, talks before work, during breaks. Many of the people outside my work group didn't know that I might be using the things they said. So I have changed all the names in the chapters about the people at work.

In the eleven chapters I have written about people's lives and their families I have used real names and photographs whenever possible. However, many of their comments about job issues, such as the bonus system, rules, and the human environment, are found in the body of material about the company, and those remarks are protected by the anonymity of made-up names.

Most, but not all, of the people interviewed were people I had worked with. I did meet a couple of people in the plant after I stopped working. Irene Lambert was one of them. She came up to me while I was photographing in her shop and asked, "What are you doing?"

"Taking pictures."

"Who gave you permission?"

"The company."

"I know that, but who do you work for? I know you don't work for the company."

"How do you know that?"

She looked up and down at my clothes—jeans and a blue work shirt, and my somewhat unkempt hair before saying, "Honey, the company doesn't have anyone who dresses like you working for them."

On the spot I asked her for an interview.

Not everyone I wanted to interview, however, agreed. Two women I had been friendly with while I worked, and who had talked a lot about their families, surprised me when they said they didn't want to be interviewed. One said, "No, don't be silly, who would be interested in an old lady like me?" I tried to tell her a lot of people would be, and then she told me quite sternly she didn't want to be interviewed.

Another woman I wanted to interview said sure, but she'd have to talk to her husband about when I could "come over." She told me later in the week that her husband had said no to the interview. He had told her work is work, and their private life is that—private—and he didn't want her to be interviewed. She said she was sorry and hoped I'd understand.

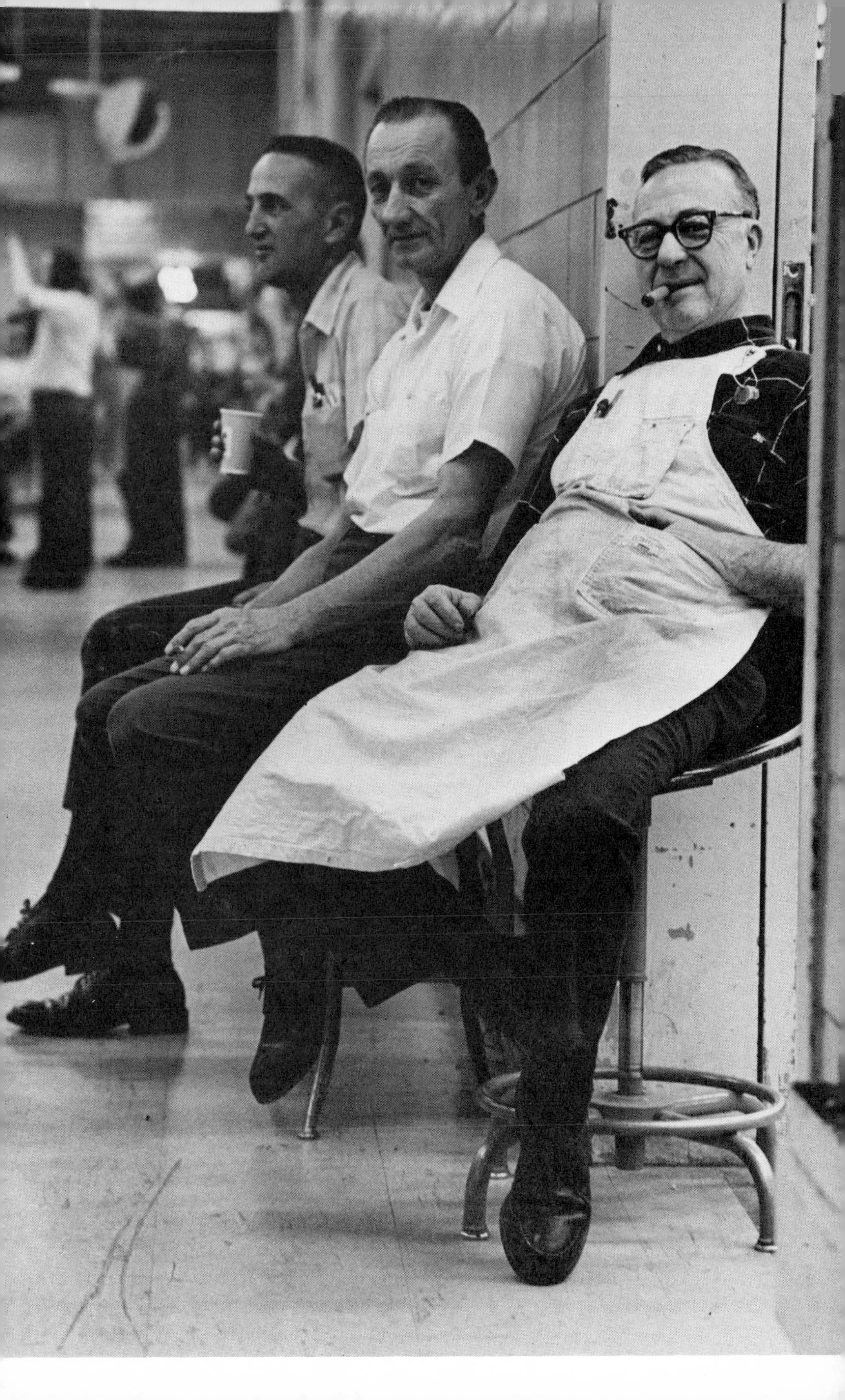

Although I didn't do any formal interviewing while I worked, each night when I went home I'd talk the day's impressions into a tape recorder. I conducted most of the interviews in homes. I began each interview by explaining that I was going to tape our conversation, that I wasn't sure what I wanted from the interview, but that when we got started we'd keep talking.

I also told people before beginning the interview that I would show them a rough draft of anything I wrote about them, and that they would have a chance to go over it to make deletions, corrections, additions, and suggestions. I wanted to involve them in more than just the taping. I wanted them to feel that they could think about the things they had said before they were published, and I wanted the stories to reflect their lives, not just one conversation.

I didn't know what to expect. The first person I interviewed called Richard Nixon a crook, and later asked me to remove that quote. Aside from that, the first two people interviewed told me they thought what I had written about them was accurate and very well done. If I had thought this was to be everyone's reaction, I was quickly disabused. A woman said, "We don't like what you've written about us. It isn't accurate; it doesn't reflect what we're like; and it's just plain boring. I'll tell you," she continued, "if the rest of your book is like this nobody is going to be interested in reading it."

Once I recovered from my initial shock, I suggested we all get together and talk about the piece. Without much enthusiasm she agreed to do that. The next Sunday my wife and I ventured up to their home for dinner and a day's conversation. We spent nearly six hours talking about what I had written. The family got me to add some things and delete others. We went over the pages again and again. We didn't resolve all the differences. No one walked away from the day perfectly pleased, but we all felt better about the piece.

Several of the people whose lives are presented in this book worked very hard at rewriting their portraits. I think in every case the portraits have been strengthened by their participation. There were things left out that I might not have left out, but I was still surprised at how open people were about themselves. Most people asked me to delete very little of what they had said. I tried to imagine how I might have reacted in a similar position. I don't think I would have been as open.

Not every interview ended agreeably. A few people had second thoughts about what they had said. Two, after long discussions, told me they thought what I had written was too personal and they didn't want their stories printed. I asked about changing names and disguising personalities, even tried it, but neither person was satisfied. Still, for a while I toyed with the idea of using the false names and then decided people had a right to veto a piece no matter how changed it was. With regret I have left those two chapters out.

A couple of other people thought I had written truthful stories about them, but for one reason or another they wanted their names kept out of the stories. I have respected those wishes.

One of the things that concerned me from the very beginning was whether I would be accepted. Many people, including some executives at the Merrimack Valley, warned me that as soon as people learned of my purpose I would be excluded, yet only once while working did I feel my acceptance in jeopardy. One morning, about two weeks after I began work, Vicky Wrigley was teaching me a new job. I looked up and saw Dave Hilder, the plant manager, and an assistant walking in our direction. I was hoping that they wouldn't stop. But they did. They said hello to me and Vicky and asked me how my work was going. No sooner had they left than other people in the group started saying things to me. There was some kidding, but there were a lot of hostile remarks such as, "I've been working here for ten years and he's [Hilder] never stopped to talk to me."

I felt a coolness from a few people that day and for several days after. I was quick to get a message to Dave Hilder that I'd appreciate it if management people on their walks around the plant would not stop and talk to me. After a week or so things appeared to return to normal.

Other than that one occasion, I never felt that being accepted was a problem. I can't be certain, but I think I was.

One day people helped me when I was violating a rule by making iced tea for several of them after the permitted break. I had left my bench position, collected the cups, filled them with water, and gotten the tea from Pauline Melling's bin. I was in the midst of making the tea when several people began to say, "Psst, psst, psst." Several women behind me were pointing their fingers in another direction. I looked over my shoulder and saw the plant manager and one of his assistants walking in our direction. I immediately jammed the cups out of sight behind a box of parts, threw the box of tea back in the bin, quickly walked back to my seat, and didn't return to the tea until they had passed out of sight.

When it was time to stop working on the first shift several people suggested that I forget about writing a book and stay on working in the department.

More than being a writer, being a photographer marked me as different at Western Electric. Most people seemed able to dismiss the fact that I was a writer, but couldn't get over the fact that I had a camera and roamed freely around the plant taking pictures.

Western Electric almost never allows an outsider to photograph in the plant. Even when politicians visit, the press are not allowed to bring their cameras along. Any photographs that are needed are taken by the company's photographer. The first few times I brought my equipment into the plant it attracted a lot of attention. People who knew about my project repeatedly told me that cameras were forbidden and advised me to hide them before I got in trouble.

Since having a camera was so special, most people never completely forgot about its presence. This made taking photographs somewhat harder. Rather than trying to be inconspicuous, I worked to make people comfortable around the camera with varying success.

I accepted that there were certain expressions impossible for me to capture, and places where my camera could not go. Words, then, would have to describe these situations. Nevertheless, I hoped to capture what life in a factory is like more vividly by photographing it.

I have no illusions about my photographs being objective, for photographic images, like words, are shaped. The photographs in this book try to document the reality I saw, to enrich the flavor the words convey.

While I worked, and afterward when working on this book, I was kidded, scolded, and offered all kinds of free advice. One day during the summer that I worked I was told, "Dick, this book better be good, because if it isn't and you've come in here during all this heat, I'm going to kick your ass." Much later after I had stopped working, when my writing wasn't going well and I had missed my third self-imposed deadline, someone said, "Jesus, what's taking you so long? I could have written the Bible by now."

Of all the things people said, two remarks remain foremost in my mind. One day, many months after I had stopped working, I was talking to Greta Barnes. I was telling her that I felt some obligation to Western Electric for allowing me to work in the factory when so many other companies had turned down my request. She fixed her eyes on me and said, "You have an obligation, and it's to us. You know, when you came to work I never thought you'd be accepted but you were, you really became one of us. You have an obligation to tell our story."

The other comment was made to me while I watched some men playing cards. An old curmudgeon asked me what I was doing taking pictures. Bill Beal said, "He's doing a book about this place."

"A book," the man said. "You ever work here?"

"Yes," I answered, "for five months."

He started laughing. "Hell," he said, "I've worked here twenty-two years and I don't know if I like the goddamn place or not, and you worked here five months and you're going to write a book about the place."

I did work only for five months and in writing this book have tried not to forget that fact. I spent nearly a year and a half after working in the plant doing more research, photographing, and interviewing. I have tried, however, to be modest in my claims to understanding.

I feel both an obligation to Western Electric for allowing me to undertake this project, and to the people I worked with for sharing their lives with me. When all other companies had said no to my project, Western Electric showed some courage in allowing me to proceed. It is unfortunate that companies should be so afraid of allowing people to undertake such research.

It is unfortunate that many companies spend vast sums advertising the products they produce, and building a corporate image, while trying to hide or at least divert attention from the workday world of the people who produce their products.

The history of labor relations in the United States in this century has been dominated by an adversary relationship, with management pitted against workers. Improved working conditions, increased wages, and better benefits have largely had to be fought for. The recognition that workers are not machines, and that labor is more than a commodity, has been slow to develop.

Western Electric allowed me to undertake this study, knowing full well that I would find things which might not enhance their public image. However, I think they decided that on the whole they had a good operation and should not be ashamed of it. I have tried to meet my obligation to the company by being as objective as I could, by reporting what I saw.

I feel an even deeper obligation to the people I worked with and got to know. They shared their lives, both in the plant and at home with me. Getting to know them has taught me that too many of the descriptions of blue collar work lack humanity, that all too frequently workers are reduced to the bend and thrust world of time study engineers. The work world I encountered was much more human and interesting.

Clockwork is a testimony to the human spirit, to the capacity of people to transform and alter an environment in which they exist. It is about the richness and diversity of human beings, about people and their search for a satisfying life. I hope it will repay the obligation I feel to the people who made it possible.

1. STARTING WORK

Before beginning work I had to fill out a job application, take a medical exam, and go through an orientation program.

I had to leave the following questions blank on the application: kind of work desired, what other work can you do, and salary or wages expected. I didn't find out until I began working a few days later that I'd be a 32 grade bench hand and would be paid $2.76 an hour plus bonus.

My job application filled out, I went to have my physical and was told I was in good health except, as the doctor put it, for being overweight. I had the entire weekend to think about the doctor's remarks and about starting work.

Even though I didn't have to report to orientation until 8:00 A.M., I got up at six, made breakfast, took a shower, and left for the plant, which was a twenty-one-mile, thirty-minute drive. I arrived about twenty minutes early and stood outside with a few other men waiting for orientation to begin.

At precisely eight o'clock the employment office doors were opened, and thirty or so new employees filed into the office to pick up their papers and passes. Because of a foul-up in the paperwork, my papers, and those of one woman, weren't ready, and we were separated from the group for twenty minutes before our papers arrived.

We rejoined the group, which was being welcomed to the plant first by Frank O'Donell, of the training organization, and then by Bill Banton, an assistant plant manager. After greeting everyone first in English and then in Spanish, Banton said, "I'd like to tell you a little about the Merrimack Valley Works. We produce more than one million dollars' worth of parts a day. At this time we have more than eight million dollars on back order and by the end of the year we will be producing at the rate of one and a half million dollars' worth of material a day. We do no military work at the Merrimack Valley Works.

"With such a big production schedule and the backlog of work we have, attendance is very important. Each and every one of you is important to keep this company functioning well, and we depend on your helping us do our job."

The theme of regular attendance was repeated first by Frank

O'Donell and toward the end of the afternoon by my new supervisor, Jack Colpepper.*

Bill Banton's remarks were followed by two films which Frank O'Donell introduced. "The first one, we hope, will explain what Western Electric does for the Bell System, what you can do for Western Electric, and what it can do for you. After that we'll take a look at a film strip about the Merrimack Works." The lights went out, the projector whirred, and the film began with the words, "The telephone is a magical device which enables a mind here to touch a mind there."

The film strips were followed by our introduction to Mr. Dan Denney, a square-shouldered, white-haired man in slacks and open-collared shirt who heads the company's counseling program. Denney explained that he wasn't a psychiatrist or psychologist, just a street person who had been around, and, as he said, "Let me tell you, there isn't anything that you could tell me that I haven't been through myself."

He urged us if we had a drinking or drug problem, or just felt hassled, to come and talk to him. He concluded by saying, "Feel free to come and see me. I hope all of you enjoy working here. The company has been good to me and if you give it a chance it will be good to you."

Dan Denney's talk was followed by a break. We all left our folders and went to the cafeteria upstairs. The men sat with the men, and the women sat with each other. At only one of the six tables we occupied were men and women sitting together.

Back in the auditorium we spent the remaining part of the morning going through the company's benefits, all contained in a thick packet each of us had been issued. There were holidays, pensions, Blue Cross, Blue Shield, major medical and life insurance to be talked about.

We broke for lunch and reassembled to be introduced to our supervisors. Each supervisor came in wearing a tie—the badge of his office—picked up his new employees, and went out into the shop. Maybe two thirds of the people had met their new supervisors when Jack Colpepper came in and picked up Mike Beal and me.

As we started out into the shops, he said, "You'll be working in an area called T2-M1, 2. Remember that name. You may get lost during the first few days, but as long as you remember what area you work in someone can help you find us. Also remember the telephone extension

* Not his real name.

(2781) and then you can pick up a phone in any part of the plant and get back here."

Jack took us walking through the plant. Its size was overwhelming. There are two million square feet of space at the Merrimack Valley Works, 593,000 square feet in the basement—referred to as the first floor by management and, in my department, 7,700 square feet. We walked past a number of groups before arriving at T2-M1, 2, marked by a little sign hung over the supervisor's desk. Jack explained that our shop was special in that it was a miscellaneous shop, that we did a variety of jobs, and that we might be asked to switch around from job to job.

"I think you'll find the work easy and you'll get the hang of it," Jack said.

"I hope so," Mike remarked. "The only job work I know is army work. I've never done this kind of work before."

"Neither have I," I quickly added.

Jack told us not to worry, assuring us that we'd get the hang of it and that we'd have a day of training before we started anything. Before bringing us back to the training area Jack made a little speech about attendance. If we weren't there regularly we'd have to be let go.

"I expect you to be in your seats ready to work at six-thirty," Jack said. "If I were you I'd give myself plenty of time. Get in the habit of getting here early, because the parking lot gets real crowded five or ten minutes before starting time. I'm sure we won't have any problems."

Jack took us back upstairs and we sat around relaxing until Frank let us go home. I got up even earlier the next morning, around 5:00, to make sure I'd get to the plant by 6:30. I arrived at 6:15 and followed several hundred people into the plant, flashing my new identification card to the guard as I passed by. I punched in for the first time, and at 6:30 a buzzer rang, people tossed their cigarettes into nearby receptacles, and went to work.

Instead of going to work, Mike and I went down to the training center and joined about a half-dozen people from the previous day's orientation program. There were film strips and practice boards. We learned rudimentary hand insertion techniques, and practiced on boards all morning. It became obvious by late morning that there was no hurry. People began relaxing and slowing down. By afternoon other people from the plant, first someone's brother, then someone's cousin, and finally someone's mother, came by to say hello.

I received a visit in the afternoon from Jack. He told me he had just come from a meeting with John Connors, then head of public relations, and that he now knew about my project and planned to treat me like any other worker. I told him that was just what I expected.

"Well," he said, "I was a little concerned at first because there are some menial tasks that go along with the job, like sweeping the floors, but John Connors assured me that whatever I would assign anyone to do you'll do too."

"That's how I want it," I replied.

"Good," Jack said. "I'll introduce you to the group sometime early next week."

Although I meant what I had said to Jack, I later came to realize that I did modify my behavior slightly. First of all, I never missed a day of work. There were a couple of mornings when I didn't particularly feel like going to work, and had I not felt a special obligation to my project I probably would have turned off the alarm and gone back to sleep. I didn't complain or speak up when something bothered me, and that isn't like me. For example, while I worked in the department I felt that Jack reacted to my presence by somewhat segregating me from other workers. I never said anything to him about it, feeling somehow it wouldn't be right.

I returned to the training center. The late morning's slow pace continued for the rest of the day, and the following day as well. It didn't seem that we were learning that much, so I wondered why a day, and in Mike's and my case, two days were being devoted to training. I decided that the training was only a smoke screen to allow new workers the luxury of easing into the work situation.

Through the training period, people were able to learn basic procedures and make mistakes in a fairly informal, unpressured setting. I know that just handling the tools I'd be working with, knowing their names and what to do with them, made me less anxious.

The third morning when I came to work I was much more relaxed and began to notice things that had totally escaped my attention the previous two days. I noticed that a tall wire fence stretched around all but the front of the plant and that I was entering the factory through Gate #9.

As soon as I entered the plant I looked upward and realized just how high the ceiling was. The floor-to-ceiling height was at least twenty feet. Looking out across the floor, I noticed that there were no windows, and that the inside of the plant looked like a giant Quonset hut.

My feet recalled the path to my department: down the first corridor, take a left down two more aisles, then right to section T2-M1, 2. I punched in and for the first time was aware of the sound of the clock stamping my card. All this made me realize how nervous I had been the first two days, and how unaware I had been of what was going on around me.

There wasn't much time to think, because Jack almost immediately started me on a job. He seated me in front of a hand-operated machine used to insert studs in a board, telling me, "This is a simple job really, all I want you to do is put these studs in these boards. There are eighteen studs that go in each board, seven vertically and eleven horizontally."

Jack proceeded to take a board and insert the eighteen studs. When he had finished he asked me if I thought I could do it. I told him I thought I could but probably not as fast as he had. "Don't worry about it," he said. "All I want you to do is to do it correctly."

Jack left and I got started. The studs which had fitted so easily into the board for Jack were reluctant to go in when I tried. The holes in the board seemed too small to accept the studs. I had constantly to refer to the sample board so as not to forget which studs went which way. At least when it was time to go to training I had memorized the pattern of vertical and horizontal insertions.

While walking to the training area I began to notice the noises in the plant. The noise level is not deafening and not nearly so loud as in a machine shop or car factory, but there are the constant sounds of small machines and air tools. We spent the day learning to gun and hand wrap things, and then we were taught the color coding system. At one point our trainer said there was an easy way to remember the color value system, which is black, brown, red, orange, yellow, green, blue, violet, gold, and white. Just remember the following sentence: Bad boys rape our young girls, but Violet goes willingly. I looked to see if anyone would visibly take offense at the remark, but no one did.

By midafternoon both Mike and I were getting bored with training and were anxious to get to work, to start our new jobs. We asked the trainer if we could go back to our group and he said, "Sure."

Almost as soon as we arrived back at T2-M1, 2, Jack began calling the other workers over to an open area. Pointing, he said, "This is Mike Beal," and then, pointing again, "and this is Dick Balzer. Dick," he said, "is going to be working here; but at the same time I want you to know that he is a writer and photographer, and he may do some writing about his experiences here. I've talked with the people in the front office and I've talked to Dick about his project. I've told him he is to be treated like everyone else. I've also told him that if any of you don't want to participate in his project he should respect your wishes."

Jack asked me if I wanted to say anything. I listed some of my previous work experiences, talked about the grant I had, explained a little bit about how I had chosen to come to work in a factory, and told them I hoped to get to know them a little better in the next several months.

Jack called the meeting to a conclusion. During the afternoon several people stopped by my workbench. One said, "If you write about this place, mention that there aren't wide enough corridors." Another commented, "Say that I'm a cog in a capitalist system."

2. THE HUMAN ENVIRONMENT

"Work," I was told, "is not something you enjoy, it's something you do." I quickly learned how ingenious people are at creating a social environment on the job. Our workday included not only the tasks we performed, but coffee breaks, lunch, coming in early, staying late, talking, teasing, joking, bitching, breaking rules, and figuring out ways to beat the system. For many people the day starts before the shift begins.

The first shift begins at 6:30, early enough, one would think, to discourage people from coming in before they have to. Surprisingly, however, by 6:15 the plant is already busy with people sitting or standing around in little pockets of two and three, talking, kidding, having a cup of coffee, or enjoying a cigarette.

I wondered why people came in early. Mike Malone[1] told me, "I don't know. I set the alarm, I get up, I don't feel much like sitting in the kitchen by myself having a cup of coffee, so I go in to work. I know a couple of guys will be coming in early and it gives us a chance to throw the bull for a while. It lets me relax before starting work."

Adrian Bly explained, "I don't like to rush. Rushing makes me nervous, so I get in early. That way I have enough time to clean up my area, to put things just the way I want them, and then I can talk with some of the other girls."

Ann Meter said, "I have to wake up early to get my husband's breakfast ready. He has to be at work at six, so I'm up, and I'd just as well come in here as stay at home."

Yet another woman remarked, "I'd rather stay at home but my ride picks me up about six so I really don't have much choice."

In the area where I worked there were several distinct groups that formed before work. They tended to separate along sex and smoking lines. Most nonsmokers met and talked within the work area, while most smokers formed in small clusters near the back wall where smoking before work was permitted.

Four men—two testers, a machine operator, and a layout—formed a group that congregated around a railing. It was to this group that I became attached, as did Arnold, another new worker. Rarely did a morning pass without some teasing. For a long time Arnold and I, as

[1] The names in this chapter have been changed.

the youngest and newest members of the group, received the brunt of the teasing. I always seemed to arrive later than the others and frequently I was met with some remark about how tired I looked. One morning John set everyone to laughing when he said to me, "Glad to see ya—when's your face going to wake up?" This remark was repeated in one form or another at least three times a week for much of my stay at Western. After a while I got used to it, expected it, and reacted to it by getting my own digs in.

Our morning conversation, especially during the summer, revolved around sports. There always seemed to be a game to talk about. The Red Sox, having a bad year, became a constant source of mockery. Next to sports our group liked to trade stories about "the service." Except for me, all the other fellows had served and they often compared experiences, talking about boot camp, lifers, and wearing the uniform.

There was some talk about problems in the plant. Occasionally someone would talk about his family. There was, however, distinctly less talk among us about families than among the groups of women I came to know.

Most of the women, unlike the men, brought their families to work with them. They always seemed to have pictures of husbands, children, and grandchildren. As we talked about sports, they talked about what their husbands were doing, the accomplishments or, for that matter, the problems of their children.

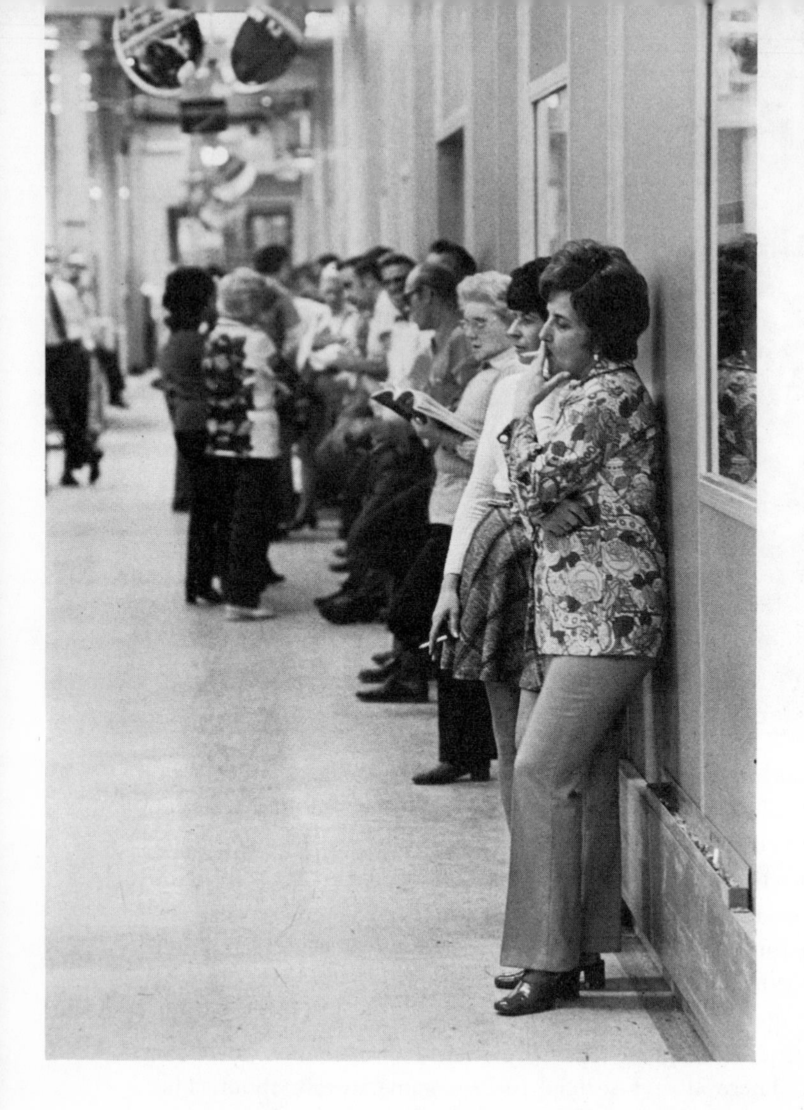

Not everyone participated in these morning activities. Some people chose not to come in early. There were a few people who, though they came in early, seemed not to participate in any groups. Content to keep their own company and wait for work to begin, they simply picked up their tools and went to their work areas.

There was one fellow who never arrived more than a minute before 6:30. When he arrived we all knew it was time to start work. The bell would ring, and, the first cup of coffee downed, the first cigarette smoked, people headed to their places.

Once the work day begins, there are three periods when people tend to congregate as they do before work. These are the two official plant-wide ten-minute breaks and lunchtime.

The groupings during the ten-minute breaks tended to be similar to the early morning gatherings. Since there were more people, the areas near the walls, the only places where one could legally smoke, became packed.

Also, the vending machines which are spread around the plant and which tended not to be very busy at other times during the day became the scenes of long lines. Many people took care of this problem by buying their coffee, tea, or soft drink just before break time. Arnold and I avoided standing on the lines twice a day by taking turns buying each other drinks.

The normal workday at Merrimack Valley is eight-and-a-half hours. You are paid for eight of those hours. The half hour is lunchtime. Crammed into the lunch periods, which are staggered between 11:00 and 12:30, people eat their lunches, talk, kid around, smoke, play cards or chess.

The people in my work area ate their meals upstairs in the air-conditioned cafeteria or downstairs where we worked. None of the men ate their meals in the work area, while nearly half the women did.

I felt somewhat lost my first week about lunch. My friend Arnold would go off to eat with his father, so this left me by myself. The first day I went up to the cafeteria nearest our area. I didn't see anyone I knew, so I ate by myself. The next day I joined two fellows from our shop. They were friendly enough, but after buying their meals and sitting down they took out a miniature chessboard and quickly became involved in a match, which I learned was an everyday lunchtime event.

The third day I decided to stay in our area and to eat with one of the women who had been helping me with my work. I went up to the cafeteria, bought a sandwich, drink, and dessert, and came back downstairs to join her and her friends. They were nice enough, but everyone was a little uncomfortable. I was the only man, and I felt like an intruder. I just wasn't a natural part of their group. Once, several months later when I knew each of them better, I had lunch with them again to see if it would be less awkward. It was but only slightly.

For a few days during the next week I ate with another new worker, Mario Caluchi, in the plant's other cafeteria, located on the other side of the building. But that did not last long. We inadvertently discovered that we were sitting at someone else's table. While we were eating, three women appeared and stood over us. They didn't say anything until I innocently asked if we could help them. They looked at each other, and one of them said, "Well, we usually sit at this table." There were other free tables, but they obviously wanted this table. We gulped down our food, apologized for sitting at their table, excused ourselves, and left. We both laughed about it—imagine people staking out a table.

Later that day during the afternoon break, I told some of the guys what had happened to us, but they didn't laugh. Eddie Larson said, "People have places where they usually eat, and that's their table."

I could see that Eddie was serious, but it didn't make sense to me, and wouldn't for nearly a month. From that day on Mario and I joined Bert, a layout from our area, and three other layouts for lunch.

Soon Arnold joined us too. Every day for four months the three of us would go up to the cafeteria together, get our food, and wind our way to the back of the cafeteria, where Bert and his friends were sitting. When our table was occupied one day, I realized that there was a pattern to where we sat. Up to then I had never thought of it as "our table," but seeing other people sitting where we sat was surprisingly disconcerting. There were other tables free, but I didn't head to another one. Instead, I stood there momentarily with the others, won-

dering why these women were using our table. Finally we sat at another table, and one of the fellows, Jerry, said, "Damn women, why don't they eat somewhere else?"

The next day we reclaimed the first seven seats at the corner table in the back of the cafeteria. Rarely was our table occupied, and each time it was, it was annoying. Once, in fact, there was no one sitting at our table, but the tables had been slightly rearranged. We actually put the table back in its regular position before we sat down to eat.

This experience made me think about the people who ate downstairs. At first I couldn't understand why so many people chose to eat by their work places, where it was hot and crowded with tools and boards, instead of eating in the air-conditioned cafeteria. I remembered (in other factories) how depressing I had found the sight of men and women eating by their machines.

I came to realize that, whether they ate at "their table" upstairs in the cafeteria or down near the machine, people were looking for personal space. Some people like to get away from their machines and work places, but others make a little "nest" of the area where they work. They personalize the area and feel very comfortable eating in that area. Those of us who ate in the cafeteria did much the same

thing. Though we went upstairs, we expected to eat in the same space, in our own seats at a particular table.

Politics, inflation, the gas crisis, and Watergate, as well as sports, were subjects often discussed at lunch. Watergate was a big topic for several months. We had a genuine interest in Watergate, but part of the reason we discussed it so often was that one of the fellows at our table was a big Nixon supporter. The guys liked to put a "burr under his saddle." When Hal would come to lunch after we were all seated, one of the guys would spot him walking toward the table and would start talking about Watergate. Several times a week, Hal would no sooner put down his lunch box than he would quickly come to Nixon's defense. We would keep talking; Hal would get madder and madder; and suddenly, when Hal seemed as though he'd explode, someone would switch the subject. None of us ever tired of getting on Hal and seeing what he would do.

Another thing we never tired of was a little game that George had invented. There was an attractive woman who ate at the far end of our table. She frequently wore tight sweaters which showed her large breasts to good advantage. When George would spot her walking toward the table, he would commence making a low guttural sound—uumph, uumph, uumph—and, as if by signal, all of us would join in. This chorus of throaty grunts was maintained as she walked by and was kept up until she took her seat. In the four months I ate upstairs she only reacted to the sounds once, when she cracked a smile. That smile was discussed for several days afterward, but she never again gave any indication that she had even heard the noises of her small cheering section.

Although Arnold, Mario, and I ate with Bert and his friends, we usually stayed in the cafeteria for less than twenty minutes. Weather permitting, we'd go outside for a smoke, walk back to our area, punch

in, and talk to a few of the women until the bell rang. Most people returned several minutes early. One took a daily ten-minute nap. Two others carried their chessboard down from lunch and continued their game in our work area.

Lunch and the two ten-minute breaks took up less than an hour of an eight-and-a-half-hour day. Many people found themselves bored with and disinterested in the tasks they performed several times during the day.

I asked Steve Diner about the clock and he said, "On really bad days it doesn't matter what you do, the day won't hurry for you. Three o'clock seems like it will never come. Thank Jesus, there aren't many of those days. I'll tell you when they come I can't stand it, it's the only time this place is really hard on you. And it isn't the work. Hell, half the time I think my twelve-year-old kid could do my job. But when the clock is fighting you it's nearly unbearable."

Neither the clock nor the job seemed to get to some people. There were a couple of women who rarely got up from their work. I asked one how often she got bored. "Oh," she said, "once in a while I get bored, if I'm not feeling good or if Jerry [her husband] and I have an argument, then things go badly, but otherwise, really, I don't mind any of the jobs. I mean, a job is a job, and I just do it. That way I find the time goes faster. It's only when you start watching the clock and ignore your work that you get bored."

I asked another woman and she said, "I don't let myself get bored. If you let yourself get bored on one job, you could go crazy with most of the work in here. I know I'm going to be in here eight hours a day, five days a week, and so I don't let it bother me. I don't kill myself. I do my job and you know, after a while, it's three o'clock."

Sometimes, however, people do get bored and they figure a variety of ways to cope with those feelings including: thinking of other things, talking, walking around, going to the bathrooms or vending machines, and telling jokes.

Ann told me, "The only boards that really bore me to tears are those dodo boards [simple boards that require no thinking]. I work as quickly as possible on those boards, just to get them out of the way and to get back to other boards that I enjoy."

JoAnne told me, "When I get bored, I think of the new swimming pool we just bought from money I'm making. I think how every hour that passes means we've paid for more of that swimming pool. I think about the pool for a while and then I feel better."

Two women who worked behind me would switch positions on a job when they got bored. One worked positions one and three, the other two and four. Then they would switch back.

Arnold told me that he would either change the number of boards he worked on or the pattern he used to work on the boards. "I have to put face plates on boards and insert nineteen pieces. Every so often I'll change the pattern. I'll work backward, and then forward. But really it doesn't help much. I hate being cooped up; sitting on my ass putting on weight doing these dumb jobs. You know, the other day I was so bored I wasn't even paying attention to what I was doing and I burned my goddamn finger."

I tried to handle my boredom by varying the number of boards I handled and the way I worked on them. I would start with one or two boards, then try to work on four boards, seeing how much of a board I could do from memory. Since I changed jobs frequently and did about ten different boards, it was a challenge to discover how much of a board with up to 120 pieces I could remember. Sometimes I'd have time contests to see how fast I could do a board.

Talking helped many people make the day go by. Our work spaces being as close as they were, it was easy to talk. A great deal of personal information was transmitted during work hours.

Much of what was said was important. However, at times it became obvious that the talk was little more than a device to pass time. Arlene told me, "When I get bored, and believe me that is quite often, I like to talk to someone. I don't think I've ever told anyone else this, but sometimes I don't even care what we talk about. I just want to hear another voice."

Ellen Smith said she didn't mind the work, but she did wish that she wasn't working in a row by herself. "It gets very lonely when I'm just working in a row all by myself. The time goes very slowly. When I have to work in a row by myself for several days I feel like screaming."

"Hell," said Ed, "if you couldn't talk and fool around you'd go crazy."

There was a lot of joking around during the day. The male layouts, who circulated freely among the bench workers, often kidded around with the women while they worked. Much of their kidding took the form of slightly off-color jokes.

One morning one of the women told Earl he looked tired. "Tired," he said, without missing a beat, "I used to be able to get it up all night and now it takes me all night to get it up." There was some embarrassed laughter. As he departed, Earl said to me, "I knew those horney witches would like that. They may look embarrassed but they love to hear that stuff."

Later that day I overheard a woman repeat another joke Bert had told her. I asked her if she was this way all the time. She said, "No, if you told my husband the way I acted in here, he wouldn't believe it, he just wouldn't believe that it was me. I'd never tell those sorts of stories outside of here. I just couldn't do it. But you know the layouts. Kidding around really helps make the time go by."

Just then Larry, another layout, came by and said, "Don't listen to her. When I came to work I thought hell was a four-letter word and now these women have corrupted me."

They argued for a few minutes about who had corrupted whom and then Larry said, "I've got a new trick to show you." He took a golf ball, bounced it on the ground and, as it rose in the air, tugged the front end of his pants forward so the ball fell between his stomach and his pants. He said, "Woo," and the ball rolled out his pant leg. It was a show stopper. The women laughed and Larry walked away with the ball.

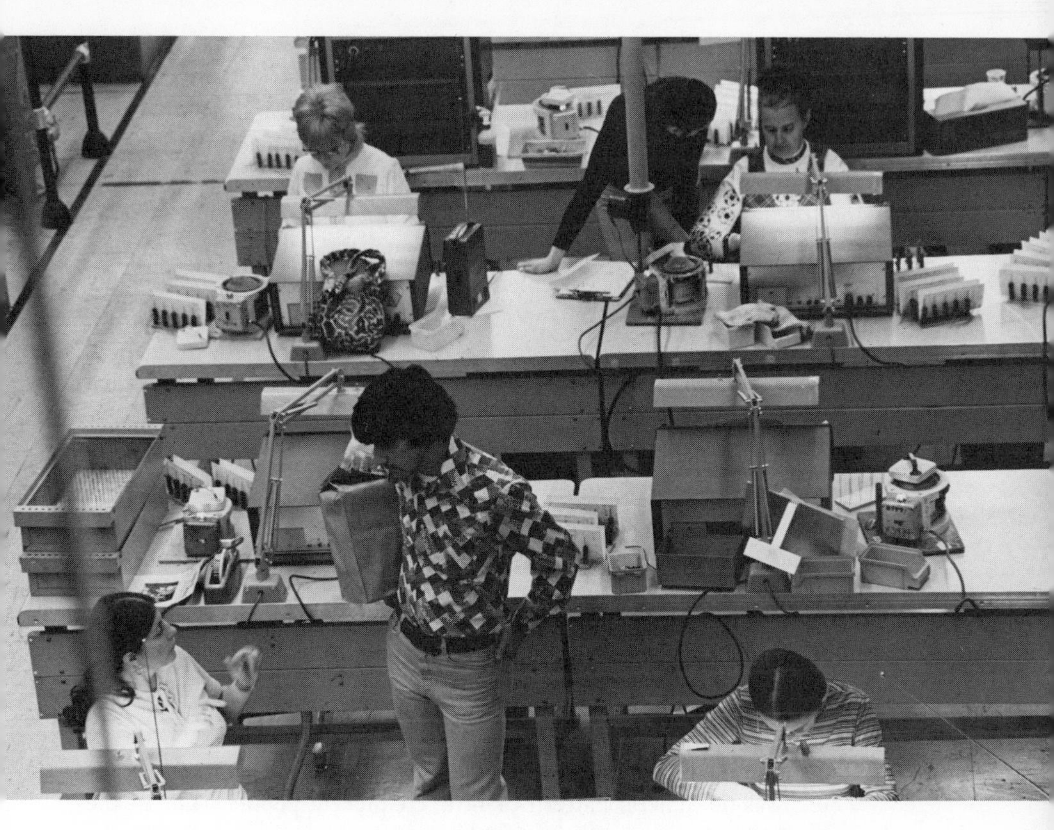

Several weeks later I asked Earl about the jokes. He said, "Look, you can't tell some of the jokes to all these broads. Some of them really would get offended. With them you just give a little compliment about the way they look or something they're wearing, it doesn't take much, and it makes them feel good. The others, they love to hear the jokes. Oh sure, in front of you or some of the other kids they pretend to be embarrassed, but they love it. I'll tell you, a little clowning around makes life more comfortable for everyone."

Often during the day, some days more than others, one feels like getting up, walking around, and talking to other people. A sense of how free you are to move around depends largely on what your immediate supervisor is like. There were some supervisors, I was told, who made you feel uncomfortable leaving your seat at all, and there were others who supposedly didn't care how much you walked around as long as you got your work done. Our supervisor gave us a certain amount of freedom, especially if we left the area.

Walking and talking in our area was something that had to be done quickly. We had the sense that Jack, our supervisor, didn't like to see us just standing around talking. We usually had time for only a few quick jokes or comments. Walking and talking almost became a game —watching with one eye for Jack while talking to someone else.

Frequently you'd be in the midst of talking with someone when Jack would approach. He didn't have to say anything. Just a sideward glance or movement in your direction was enough to get you moving.

Some people were more adept at walking and talking than others. One woman in our area seemed expert at the subtleties involved, another grossly ignorant. The expert would come by and talk, but she always seemed to have an acute sense of how far she could push, how much she could get away with. She rarely, if ever, was reprimanded by the supervisor.

The other woman moved as quickly as a giant Galápagos turtle. She'd make you feel uncomfortable by standing around talking and paying no attention to Jack's movements until she was caught. After Jack had chased her back to her bench, she'd complain that she was the only one he ever reprimanded.

We didn't have the problem of Jack's watchful eye if we left the area, so a trip to the vending machine or the bathroom was often relaxing. The bathrooms were not just places to take care of biological needs, but places to relax, either to get away from work, to enjoy a smoke, or to hang out.

There was no general pattern of how frequently people went to the bathroom. Some people seemed able to stay at their work for the whole day, aside from official breaks. Others seemed to leave the area several times during the day. The bathrooms tended to be most crowded at four times during the day: before the two breaks, before lunch, and before the end of the day. One woman told me, "They don't give you very long for lunch or the breaks, so I figure I'll wash my hands and freshen up on company time."

Although there was no discernible pattern to how often people used the bathrooms, I did notice a difference between men and women. Men tended to go to the bathrooms by themselves. For example, I can't remember ever asking one of my friends to go to the bathroom with me. It just didn't cross my mind to do it. Furthermore it didn't usually matter to me who was in the bathroom. If there were someone I knew, I'd take a smoke with him and talk a little. If alone, I was usually carrying a paperback in my back pocket and would find a stall to read in for a few minutes. Occasionally I'd stand around with some guys I didn't know, bitching about this or that.

Women, on the other hand, tended to go to the bathroom in pairs. Over a one-month period I became aware of at least five distinct pairs of women from our area who seemed to head toward the bathroom together, purses in hand.

I became more and more curious as to why this was the case. I asked several of the women why they did it. Most seemed surprised that I had noticed the practice and had no real explanations for it. One woman looked at me as if I were a pervert. "Don't be a naughty boy," was all she said. Another woman said she had never really thought about it. She just went to the bathroom with her friend because they talked a little while they washed up or while freshening up for lunch.

Denise James told me she'd think about it, and later in the day told me, "I really don't know. I think it may be something of not feeling vulnerable. I know whenever we go out with a couple or couples to a restaurant or a club, I always ask the other girls at the table if they want to go to the powder room. I feel better to walk with someone else. I guess I never think about it, but I feel somehow less exposed walking with one of my friends to the bathroom."

Talking on the phone, though not very time-consuming, also played an important role in taking people away from their work. In most of the factories I've visited, phones are not one of the amenities immediately available to workers. If there are phones on the shop floor, they are usually pay phones, often at quite a distance from the work area. At Western there are pay phones, there are also phones by the supervisor's desk. Incoming calls, particularly the details of who received them and from whom they were, were of great interest to many people. If Jack was away from his desk and the phone rang, one of us would answer, "Department 81770." If the call wasn't for Jack but for someone working on the bench, the fact was usually communicated in a loud voice. People were very quiet when the person first picked up the phone. Once the call was completed and it was clear it hadn't brought bad news, there would be a lot of teasing. One girl who was getting some ribbing and being asked several times who called said, "It was an obscene caller—it was my husband."

There were a few days in August when people looked up every time the phone rang. Ann's daughter was expecting, and the call from the hospital would mean that Ann's first grandchild had been born. Those days we waited for the call, waited so we could celebrate. By the time the call finally came, Ann had already left early, the tension too much for her. We celebrated without her, and again the next morning with her when she arrived, with cigars and candy.

Occasionally the phone would be used for a prank. Someone would call from the back of the room and a person not expecting a call would get all excited until she found out that someone was playing a joke. I was involved in a phone prank. One day, Loraine called me back to her desk. She told me a woman from another department was going to be transferred into our department but it had been called off. She said, "We want to play a joke on her. Call her and tell her you're from the union and that she's going to be transferred."

It had been a slow morning, so I didn't need much coaxing to play along. They gave me her extension. She was in the next department, no more than a hundred yards from where we were. I made the call, claimed to be a union official named Ralph Turner, and convinced her that the transfer was on again. For a last flourish I told her we were doing this all on the QT so she shouldn't tell anybody until we had it "locked up." She told me, "Ralph, I've always supported the union, you don't have to worry. I won't mention it to anyone."

Within five minutes she had come over to the three women who had put me up to it and told them she had been informed by sources she couldn't mention that she was going to be transferred. They waited several days before letting her in on the joke.

The only thing that would take everyone in our area away from work was a meeting. Meetings, though infrequently held, were enjoyed by most people. The time we spent at meetings was time we could write off on our bogies so it didn't hurt our bonuses. Therefore, no one except our supervisor was ever in a hurry at meeting time. From the moment we left our tools and benches the pace was slow. People would talk to friends in other departments as we meandered to the meeting place. You could smoke at a meeting, so we did. If there had been a popcorn concession, someone could have made a fortune.

Except when it meant exposing themselves, people would try to think of ways to extend the meeting when it was coming to a close.

The longer the meeting, the more time we were legitimately away from our work. On more than one occasion I heard people whisper, "Ask something, ask something."

Questioned out, the meeting would end, and it would be time to return to work. The leisurely pace of the walk back to our shop was in marked contrast to the desperate pace set when we left the plant at the end of the day.

The summer's heat, which was often in the nineties inside the plant, led to some special time-consuming, heat-forgetting practices. Ann, who worked in the row behind me, would appear on a hot afternoon, her hands full of wet towels, and either hand you a towel or, if she knew you better, slap it on the back of your neck. The wet towel stayed cool long enough for a few brushes of the neck, a swipe or two over the face, and a quick pass over sweaty arms and hands.

Distributing towels gave Ann a break from her work and provided others a nice break on a hot day. So did the iced tea we made each day. One of the younger women brought a big jar of iced tea mix in to work. She made five cups of tea every afternoon, after lunch and before the afternoon break.

The last of the day's activities in the plant is leaving. The three-

o'clock bell rings, and there is a mad dash for the time clock, the door, and the waiting cars.

At the end of my first days in the plant I joined the rushing throngs. I lined up near the time clock, punched out as soon as possible, and made for the exit, nearly running till I hit the cement on the other side of the guard post. Then I sprinted to my car, opened the windows to let the summer heat out, started the engine, and headed out. I never seemed able to beat the bumper-to-bumper traffic, and it was always a good fifteen minutes before I left the lot, by which time my armpits were sweat stained.

In less than a week I decided to bring in a book or magazine and read for a while before going out to the parking lot. I wondered why people made such a mad dash to get out of the plant when they knew they were destined to be stuck in long, unpleasant lines of cars. People offered a number of explanations, reflecting a deep feeling of relief when a day of work is finished, and an anxiousness to leave the plant. Whatever the reason, once that bell rang, people wanted to get out of the plant no matter what.

For many workers the company-connected part of the day does not end with the rush to the parking lot. More than three thousand employees participate in the more than two dozen, company-sponsored afterhours activities, which run from a full schedule of men's and women's athletic leagues to macrame clubs.

I played in the soccer league. Unfortunately, my team, the Panthers, had the distinction of winning only one of the twenty-three games we played. Nevertheless, I enjoyed the games and the stop off at a local club for a couple of beers afterward. Some of the players on the better teams took the games and the league very seriously. They might think of staying out, missing a day of work, but they wouldn't miss a day when they had a league match afterwards.

We had one outstanding soccer player in our group, and players from other teams were always stopping by to talk with him. One of the layouts was constantly kidding him about the number of engineers and big shots who stopped by to talk about soccer with him. "Hell," he said, "Mario's only been here a year, and he already knows more big shots than I do."

The afterhours activities bind workers sharing similar interests together. People really look forward to their group's meetings or league play.

The human environment at work, supportive for many, can be cruel and devastating for others. People can be very nice and very compassionate. They can also be very mean and catty. Many people enjoy the informal teasing, the constant joking, but others don't. One woman on the evening shift warned me, "Watch out. People will smile at you, but they'll talk behind your back. If you're smart you won't tell them anything, unless you want everyone to know it."

After working downstairs for a while I learned that workers have mixed feelings about other workers. One day in sheer frustration Ann told me, "It kills me. We bust our butts to really produce, and it's not fair that Sylvia sits there and doesn't do anything. My husband tells me that's how it is. He worked on a farm, and he said the strong horse always pulled extra for the weaker horse, and that's how it is in here."

Janie Silver told me she didn't enjoy working near a certain woman because it made her nervous–you didn't know when she was going to throw something or say something, or tear into somebody.

Still another woman told me, "Judy knows I'm friendly with Claire, all the girls know it. I know they don't like her but why do they talk about her to me? They want me to say something bad about her, but I won't do it."

Generally, these negative comments remain fairly muted, but the kidding can occasionally cause serious unhappiness. Tempers can flare and a person can start crying. One day Bertha told me, "I was just joking around with Jane. It was so hot I said her back must be sticking to her chair. Well, she got all huffy and told me she didn't like that talk. She yelled at me. You know what I did? I didn't say anything, I just walked into the bathroom and cried. Why'd she talk to me like that? I was just kidding around with her. We haven't said a word to each other for the last two weeks."

One day in September I was working on some "296" boards when a little teasing erupted into yelling. All of a sudden, without warning, Janet started yelling at the top of her voice, "I won't have it, I won't have it." She slammed down some boards and started to cry. She sat at her desk for more than a minute crying, her face buried in her folded arms, before a friend came over and took her away to one of the bathrooms. The two women who had been teasing her couldn't understand. "We were just joking around," one said. When Janet came back from the bathroom she said, "I don't know, it just got to be too much for me. I just couldn't take it anymore."

Most times the teasing comments are innocent—the intention of the person making them is to have a little fun but not to drive someone to tears. People are sometimes intentionally cruel. Their comments are meant to hurt, and they do. The most graphic example I remember was on the evening shift. A woman, who had just transferred to the department I was working in after being at Western for several years, had spoiled a number of boards. The layout, a woman, really chewed her out. Dale didn't say anything in her own defense. She just sat there as the woman yelled at her. When the woman left, she began crying. She continued to do her work as the tears ran down her cheeks. I went over and tried to comfort her. She kept wiping away the tears, but they kept returning. In a halting voice she said, "What did she think, I wanted to make those mistakes? I didn't know how to do the job, no one really taught me. I didn't want to do it wrong." Just as she said, "Why'd she talk in that horrible way to me," she began to cry again, her mouth curled, and she couldn't talk. I told her we could discuss it later and suggested she go to the bathroom to calm down and to pull herself together.

When she came back, she said she felt better. For the rest of the night she cried occasionally and talked about Nancy. "That's what I can't stand about this place, the way some people treat you. Nancy treated me like a dog. I'll get her back, you just watch, I'll get even with her."

Just as the teasing can go too far, some people resent curiosity others show about their lives. June Bycyk told me, "I don't like the way these women are always prying into your life. They want to know everything about you. If you talk to a man in here, they think there's something going on. I'd rather be by myself than spend time with a lot of them."

Some people choose to limit severely their participation with the group. I've seen persons who preferred to sit with a book by themselves during a break than to chew the fat with other people.

There are, unfortunately, some people who would like to participate but who are ostracized. One fellow on the evening shift comes to mind. He was physically and socially awkward and became the butt of many jokes, some good-natured, others not. He tried to eat with a group of us on a couple of occasions, but when he sat down the other guys either ignored him or stopped the joking and laughing. I asked one of them why this happened, why people weren't friendlier with Stan. He said, "I don't know, man, he just is a creepy dude. I just don't enjoy eating my dinner with him."

Stan was consistently unsuccessful in his attempts to be part of the group. At dinnertime Stan would either eat by himself in another part of the cafeteria or would go see a "friend" elsewhere in the plant.

Stan's treatment reflected the more noticeable tension and hostility I found upstairs at night. In a month upstairs I saw people cry, challenge each other to a fight, and heard a number of stories about this one and that one. The ugliest scene I witnessed started innocently enough. A young fellow knocked a table that a woman was working on. It was something he often did. She started screaming at him. As he walked away, she hit him. He swung around and threatened her, saying, "Don't you ever do that again because the next time you do, I'll put you through the wall." Nothing further happened but neither talked to each other for a long time after that.

After that incident another worker in the group told me, "There's a lot of backbiting around here. One day someone will tell you how much they like you and the next day they will be talking about you behind your back, and saying what a shit you are. It can be real ugly in here."

Whether good or bad, the human environment is important to most workers. For some it can be supportive, allowing people to connect pleasurably with each other. For others it can be cruel and punishing. It can help bring some people out and force others to withdraw from social interaction, to isolate themselves. Good or bad, the human environment defines a major portion of a person's working day.

3. STOKING THE HOME FIRES

VICKY WRIGLEY—In the T2-M1, 2 shop each new worker was shown his job by a more experienced worker. Vicky Wrigley became my teacher. She not only taught me how to do the jobs according to the visual aid supplied by the company, but also showed me the short cuts and little tricks that an experienced worker knows. When I had trouble she often said, "Come tell Mother about it."

It was a hot, muggy July night, the kind that makes a cotton dress cling to the body, when Vicky Walent met Harry Wrigley. "They used to have open air dancing," Vicky says, "and all the girls used to like to go together because no one wanted to dance with the same boy all night.

"Harry was at the dance, but I didn't want to go out with him. One of the girls told me, 'Every time you see him, he's with a different girl.' So when Harry came over and asked me, 'Would you like a ride home?' I told him, 'No, thanks, I'm going home with my friend Ginger.' He didn't say anything, he just walked away.

"Well, about a quarter of twelve I started looking for Ginger to go home, but I couldn't find her. So my friend over here [Harry] came over and innocently asked, 'Are you looking for somebody?' He knew I was looking for Ginger.

"When I told him Ginger was nowhere around he offered me a ride again. I didn't know then that he had schemed with Ginger, and told her she didn't need to stay, because he was going to take me home. Anyway, I was stuck at the dance and scared to death to go home with Harry."

"I took you right home, didn't I?" said Harry.

"Yes," Vicky laughs, "you did."

"I didn't even make a pass or nothing."

"No, but the next night you came banging on my door asking would I like to go swimming. I told him no but he said some of my friends were in the car and that I could trust him.

"I went swimming and we were married a little more than a year later, and that was nearly thirty-five years ago.

"That's a long time ago," Vicky says, "but the time has passed pretty quickly. When I think about it, I've been working at Western Electric for nearly eighteen years. When I started I thought I'd only work a couple of years. But every time I thought about quitting we always seemed to need something else, so I kept on working.

"Once, several years ago, I had thought about quitting, but Harry was unhappy at his job as an insurance salesman. I told him if he didn't like the work he should quit while I was working and do something he liked. He did quit and he started selling automobiles, and he's been selling cars ever since.

"Thinking back like this makes me think about when the children were small, and how we used to play leapfrog with them on the carpet." The children still take up a large portion of their parents' time. Rusty, their son and the older of the two children, is a teacher at a local secondary school. He wanted to be a teacher for many years, and now he wants to be an administrator. He has his master's degree. Vicky says, "We really don't help him out financially anymore. Sure, I might give him a little something to help him out with his books, and I can't help but spoil my grandchild a little, but Rusty takes care of things."

Harry says, "He's the kind of kid that a mother likes. He's dependable and solid. Rusty's the typical guy that mothers like. They go to school, get a job, get married, and have children. Now Rusty could have gone completely the other way and become a gypsy, but he didn't."

"That's right," says Vicky. "He married his high school sweetheart. But he always had sense. They waited until they each graduated from college before they got married."

"See," says Harry, "that's just what I mean. He and Terry loved each other, but they waited to get married until after college. You know, just what mothers want. Nice wedding, big reception."

"Someone will ask me," Vicky says, " 'What does your son do?' I'll say, 'Oh, he's a teacher and he has his master's.' Then they'll ask, 'What does your daughter do?' I'll start, 'She's . . .' and then I slur a little. I mean that's how things are. Joannie's a good kid. I can't say anything against her. Whatever she does, she does on her own and she enjoys what she's doing.

"When she comes home, she loves to be home. She's my daughter and I enjoy her. I'd love for her to be living here, get up, go to work, do her job, come home, but that's another thing. That isn't her way."

"Do you think that Rusty going to school locally and Joannie going to school in Colorado made a big difference?" I ask.

"No," says Harry, "they've just got different personalities. Rusty is a homebody. He wants to lay down the rules and make plans. First it's I'm going to finish school, then get married, and then pay for a house and have children. Joannie could decide tomorrow she wants to go to Alaska and she'd just pack up and go. Rusty, if he was going to Alaska, it would take a year's planning to do it. Joannie could hop in the car with a dollar in her pocket and say 'I'm going to Alaska' and she'd go. I envy Joannie for that."

"Joannie's never had a real plan of what she'd like to do," Vicky says. "Joannie went to a junior college around here for a year, but she didn't like it. At the end of the year she told me she wanted to take a year off. Well, I wanted her to get an education. I never had the chance to get one, and I always regretted it. When I was young I thought about becoming a nurse, but I knew going to college was out of the question so I never told my parents. We didn't have the money, and besides girls didn't go to college in those days. The closest I ever came to college was taking some clerical courses when I worked in Andover. I never used them, but I was glad I had the courses. I wanted my children to go to college, if they had the chance, and we could help them out. Harry and I have always tried to help them out if we could.

"But Joannie didn't like school and wanted to work, so she got herself a job in a local factory. I remember I warned her that she might not like it very much. We didn't argue about it. She just told me she could handle it.

"After only one day at the rubber plant she came home exhausted. She came in the front door, all tired out, and starting laughing, 'It's just like you said it would be, Mom. I never thought it would be like that.'

"When I woke her up the next morning for work, she told me she didn't feel too good and maybe she'd stay home. I told her, 'Look, dear, you're working now. If you only have a cold you should go to work.' She lay there for a while and I told her, 'All right, young lady, get out of that bed and get going.'

"I had to tell her the same thing the third morning, and when on the fourth morning she told me she didn't feel well, I let her stay home. She quit the job that Friday.

"I'll say this for her, she didn't lay around that summer. The next week she went looking for a job, and started working for an insurance company in Andover. I think that little taste of what work was like convinced her that going to school was a better idea. She enrolled the next year at the University of Northern Colorado and she just graduated last June.

"I took ten days of my three-week summer vacation to go out to Colorado to pick Joannie up. She wanted me to meet her friends and then we were going to drive around. She was going to show me Utah, the Great Tetons, parts of the country I've never seen. But Joannie's car had problems, one of the valves had a pretty low compression reading, so we came home as quickly as possible.

"The trip gave us a chance to talk. Joannie had all sorts of ideas about going to South America or India. I told her after the last tuition check she would have to start assuming responsibility for things. We would help her out, but she would have to find a job, and begin supporting herself.

"I promised her a trip as a graduation present. Well, she asked me for the money, saying she might use it to go down to Latin America. I wanted to give her a nice comfortable trip as a present and I thought if what she wanted to do was to go to Latin American she should do it by putting her feet down on solid ground first.

"She said if I'm going to give her the money why should I care how she spends it? I guess she's right, but I do worry about her. I think we're very close, but in some ways we're very different. Last time Joannie was here I drove her to the Motor Vehicle Bureau to have her picture taken for her new license. Her hair wasn't fixed up, and I said, 'You know, you look like a washerwoman. Aren't you going to make your hair up?' Joannie just looked at me and said, 'No, what's the difference?' I told her the picture had to last four years. 'So what?' was her answer. 'It doesn't matter, it's not important to me. It's just a picture of me.' I told her if it was me I would have been sure to fix my hair up.

"I know with a lot of these things that sometimes I hold on too much. It's hard to let go. I know she's twenty-four and can make her own decisions, but still, it's hard to let go. I've been that way for a long time, so the kids know that when I said no they should ask Harry.

"I remember when Joannie was eighteen and she wanted to go to Canada with a couple of friends. I wasn't too anxious for her to go, but Harry told me, 'Look, she's a good driver, let her go!' Finally I said OK.

"She called up from Montreal the next day and said she'd been in an accident. I was so excited when she said that that she asked to speak to her father. She told him about the accident and that she had already called the insurance agent, and had had the car repaired. You know what Harry told her, 'Now I want you to find the nicest restaurant you can, have the best meal, and then find a nice motel and stay there.' Well, she felt just like aces. Harry didn't criticize her; he was proud that she had taken care of things.

"A few years later she wanted to go with her girl friends to Alaska. I was hesitant again but Harry said, 'Look, she's twenty-one now, it's time we give her our blessings and tell her to do the things that she wants to do. We can't stop her anymore!'

"So Harry is more lenient with the kids. I think some of it is because of what it was like when I was a kid. I grew up in a small town in Maine—Mexico, Maine. My mother was always saying no; she'd say no before you'd ask. It seemed as though I could never make plans and say, 'Oh yes, my mother will let me go.' I could never have that freedom—never.

"When I was in high school I was involved in lots of things. I was the head cheerleader, and I was on the debating team. If you were on the debating team it meant you had to do research in the library. I'd say to my mother, 'I'd like to go to the library tonight.' She'd say, 'No, you're not going to the library tonight.' Or I'd want to go to a game or practice for a game, and she'd say no and she'd never explain the reason.

"After my mom would say no, I'd ask my father, because I knew he'd usually say yes. Usually he'd say something like, 'OK, but come right back when it's over so your mom won't be too mad at either of us.'

"My dad died thirty years ago and that left my mother, and bless her soul, I love her, but she's not always that easy to live with. She doesn't like to let go. I'm almost embarrassed to say I still wonder before I do something whether or not my mother would let me do it. Imagine that, me, at my age, and I still want my mother to say OK. It's only been the past couple of years that I've really started to do the things I want. I think it started with our trip to Hawaii. My sister and I planned a trip and I told my mother, 'We're going to Hawaii.'

"She said, 'I don't know why . . . there's plenty to see around here.'

"It bothered me a lot, but we finally went.

"So even though I say I won't hold on like my mother, I know I hold on, a little more than I'd like.

"Joannie's a big girl now, and I still worry about her. I'd like her to be a little more financially secure. She's happy, I know she's happy, but I still worry about her. She's living in New Hampshire now and a boy she was going with in Colorado has moved to New Hampshire.

"I don't ask about certain parts of her life; I try not to put my nose into certain parts of her life."

"It's the old ostrich syndrome," Harry says. "Put your head in the sand and you don't know what's going on. Times have changed, and things are different. When we went to visit your parents before we were married, we did our thing, necking, in the automobile."

"We never kissed in their house," Vicky says. "We would never even sit next to each other. I wouldn't dare to kiss you in front of my parents. Kids today are more free. We aren't puritans, but everything has its proper place.

"I know my thinking is affected by the way I was brought up," Vicky says. "Just like in certain ways I still have a problem spending money. I was a kid right after the Depression. My mother needed every bit of money my father made just to clothe us four kids, to feed us and keep us warm. There were two things I wanted as a child and never could have. I never had a bicycle to ride; I never had a doll carriage. The only way I had a doll was when you bought bread. There were coupons; save so many coupons and get a doll. That was the only doll I remember having. We kids would have to entertain each other. Our big fun was that somehow my father would take us for a ride. My father bought a car just before the Depression. Somehow he kept up the payments and he would have a little money, maybe a dollar, to put some gas in the car and go for a drive. He'd always have a couple of nickels in his pocket so the kids could have candy.

"My other great joy was picking fruit. There were some farms nearby and for fifty cents you could pick a huge barrel of apples or plums. My mother would can the fruit and put it away for the winter. You know, there's still a farm nearby where you can pick blueberries. Harry doesn't like to do it, but I do. Instead of canning, I just put them in the freezer and use the blueberries for pies.

"I still have trouble spending money and my mother never wants us to spend money. Just last summer I thought of renting a place near the beach for my vacation where I could take my mother. I told her about it and she asked me, 'Still got the mortgage, don't you?' I told her yes and she said, 'Well, put the money away toward the mortgage.' Now my mother knows I've never missed a payment on the mortgage and I never will. Yet she doesn't want me to spend money.

"For my mother some of the nicest Christmases were when she had enough money to pay off some bill she owed at local stores. So she doesn't like me to have any outstanding bills. But things have changed, Harry and I have been able to work and make a good life. For years things were tight. We moved into our first house thirty-one years ago. Just when we had nearly paid off the mortgage, Harry decided to buy the land where our house is now sitting. He came in one day and said, 'Phil's decided to sell us that piece of land.' I was worried about it, but Harry thought we had almost paid off the last house and it was time to move. Harry helped build the house we wanted; he helped design it, and build it; within a year we moved in. Just when we were getting out from under we had another big mortgage. Then we've had the children's educations. That was something we wanted to do, but it was a major expense. It's only recently, within the last five years, that we could afford certain things we wanted.

"You know, I only got an automatic washer about five years ago and I still have the old washer in the basement. I know I should give it away, but who wants it? Nobody wants to use a hand washer anymore. Even people on welfare want an automatic washer. It's just like a black and white TV. People on welfare don't want a black and white TV; they want a color TV. We got a color TV three years ago. I told Harry, 'Look, let's treat ourselves well; let's get a color TV.' He said naw, anything he wanted to watch he could watch on black and white as well as color and I coaxed him into it. Wouldn't you know it, we got it and Harry wants to watch all his shows in color and I end up watching the black and white when I'm downstairs ironing.

"The trouble is we still have a hard time spending money. I guess it still makes me a little uncomfortable."

Harry agreed, "We were brought up to watch a dollar. We have a few dollars now but still we can't throw it away without feeling guilty. If we were to spend a hundred dollars today it wouldn't hurt us, we wouldn't miss it, but we'd think twice. We could spend a lot more money and not go without, but you just don't do it because you have it. I think that's the times we were brought up in."

Vicky says, "I'm satisfied that I have helped our two children. They wanted to go to school and we were able to help them. This gives me some inner peace. Now Harry and I should spend some on ourselves, but it's difficult."

Vicky gets up to go into the kitchen and says, "I'm going to put a couple of baked potatoes in. Please stay for dinner; we're having roast pork. . . . Oh," she says, "that's right–you're Jewish and I'm asking you if you want pork. You don't follow that, do you? If you want, I'll put a steak on for you."

"No, that's fine. I eat pork at home."

"Really, Dick, if you don't like it, I'll give you anything you want."

"Thanks, Vicky, it will be fine."

A little later Vicky yells from the kitchen, "Come to the table. Dinner's ready."

We dig into the meal, the fire still crackling in the background. Harry and I each take a new beer as Vicky says, "Harry's been here all his life, but I was brought up in Maine. I didn't come down here till I graduated from high school.

"There wasn't much work in Maine so I came down to Andover, where my sister was working. I worked in the comptroller's household at Andover Academy. I did cooking and taking care of laundry. It makes me laugh when I think how hard I worked the two years I was there. I worked incredibly hard, and I can't say I liked it, but it was a paying job.

"I enjoyed my next job more; I worked for a very wealthy woman who had a home in New York and a five-story summer home in this area. I worked hard. The woman treated me very well, almost like a daughter. I used to be friendly with her daughter.

"I remember once after I stopped working the girl invited me to visit her in New York. I didn't go, because I thought I'd have to have all kinds of fancy clothes, or I'd be embarrassed. I didn't keep up with her. I think now I would have acted differently and kept up with her. Now I can see how silly I was. I know if my Joannie was in a similar situation, she would act differently. That's a generation difference. When I was young, if you were out of your group you couldn't be satisfied as to who you were. Even though this girl was willing to accept me, I would have wanted to have more. I don't think Joannie thinks that way. She wants people to accept her for who she is, and not what she has."

4. THREE BROTHERS

The Cascio Brothers—Gandolfo Cascio was introduced to me as Andy. I called him that for several months until I realized everyone who played soccer with him call him Gandolfo. His brothers Matteo and Giuseppe (called Joe by nearly everyone) frequently came over during the ten-minute plant-wide afternoon break to chat and have a smoke. They'd talk in Italian unless Mike Beal or I joined them.

On the evening of November 7, 1973, the three Cascio brothers dominated the second annual Western Electric soccer award banquet by collecting four trophies and two championship jackets.

North Andover is a long way from the small Sicilian town of Polizzi Generosa, where the brothers were born. Gandolfo, the first born (1939) was named after the town saint. He was followed by the only girl in the family, Pietrina, in 1941. Then came Matteo in 1943, followed by Domenico in 1946, and Giuseppe in 1948.

As the family grew in size, the economic pressure increased on the father, Bartelo, who ran an ice business like his father before him. So at fourteen Gandolfo stopped going to school and began to work. He got up each morning before five and rode his bike fifteen miles to a construction site.

Gandolfo remembers those times. "Don't think I wanted to quit school. I didn't. My father needed my help, and so school was through. I'm not complaining, that's how life is. I was a big strong kid and I didn't mind the work. My problem in those days was my temper. What do you think? I was only a kid, and I fought a lot. Especially I was the oldest of the family and I had to protect my family."

Construction work in Italy is seasonal. The steadiest work is with the government. When he was seventeen, Gandolfo was able, with some help from his father, to land a job with a government firm. By then he had replaced his bicycle with a motorbike, and for a while work went well. Then Gandolfo lost his job. Gandolfo doesn't talk about it, but Matteo says that his brother lost this job because along with several other young men he tried to organize the workers. Although there was no strike, and the demands were met, those who organized were let go.

Unable to get unemployment compensation, Gandolfo began to think about working abroad. "I thought about going to Germany, but at the time Germany and Italy had an agreement that for an Italian to go to Germany he had to go with an Italian contractor. So my cousin and I signed up with a local contractor. We had to agree to work for the contractor for a year, and then we were free to look for work ourselves." In April of 1960 Gandolfo Cascio was the first of the children to leave Italy.

"For me, the work was not that bad. It was like being in the army. We all worked together and lived in barracks. We made less than free laborers and maybe we had to work harder, but it was steady employment." Gandolfo sent money home, and after a year he began writing about the possibilities of Matteo coming to work. He contacted his boss and asked if he would take Matteo on. The contractor agreed and Matteo, who was excited about joining his brother and going to Germany, made arrangements to join him.

Matteo says, "Up to then I didn't really work. Sure, I quit school at fifteen, and I did construction work, but I worked for a construction outfit where my father was boss. I didn't get paid much, about eighty cents a day, but I didn't work much. My job was to bring water around for people. When I think of working hard," Matteo continues, "I think of my father. He worked Sundays, holidays, and vacations.

"When I went to Germany I had to sign a contract like my brother, but I was luckier than he was. Gandolfo had already learned German, and so when I moved in with him it was easier for me. Still, for a while I felt awkward, but as a young man of eighteen or nineteen you can pick up a language pretty good.

"When I first arrived I noticed there was something wrong with my brother's chest, but he never wrote home to the family about it. You can't see it anymore but his chest was all funny. He was working down in one of the pits, putting in pipes, and the hole collapsed around him. He was lucky to jump high enough so his head wasn't buried. But the rest of him was buried and he badly injured his chest.

"I never liked the work. After six or seven months I told my brother I didn't like it and I wanted to leave. He was a good worker and had been given extra responsibility, so he was less anxious to leave. Finally we both took jobs at Mercedes-Benz.

"We had made two marks sixty pfennings when we left the construction job. We began at Mercedes-Benz at three marks and eighty-five pfennings an hour and rose to five marks fifty-two pfennings before we left. We were the second highest paid group in the plant. It was a good job. I still can't figure out how you say it in American, but in Italian we call it *carrozziere.* It means we put on the doors and different parts of the car.

"We moved into a place in Sindelfingen near the factory. From the time I came to work in Germany, Gandolfo and I worked in the same place and the same department for twelve years. It's only since we both started at Western Electric that we worked in different departments.

"At first," Matteo continues, "we sent money home every month, but after a while we sent money less frequently."

Gandolfo says, "I'm not embarrassed to say we sent some money home. When all the kids started growing up and my father needed the money less we sent less. After a while we only sent money home for holidays."

Matteo laughs and adds, "I think my father just wanted to make sure we weren't going to start asking him for money."

While the two boys worked in Germany, their sister Pietrina had married and moved to America. Dominick (Domenico's Anglicized name) was growing up. Soon after he turned seventeen, he thought of joining his brothers in Germany. By the time he was ready to go, the laws between Germany and Italy had changed and an Italian didn't need a work contract. Dominick came to Germany with his father, who had decided it was time to visit his two eldest sons. The father stayed for fifteen days. It was a pleasant time for everyone.

Before Dominick's arrival, Gandolfo and Matteo talked about where he should live. Both agreed that he should stay with one of them until he became familiar with the area. Matteo decided to move out.

"I didn't want to move," says Matteo, "but it was best for Dominick to live with one of us, so I moved in with some other friends."

Both Gandolfo and Matteo wanted Dominick to join them at Mercedes-Benz, but the company had a rule that no one could begin work until he was eighteen. Matteo made arrangements with some people he knew in personnel to take care of the paper work early, so that Dominick could begin work as soon as he turned eighteen.

Back in Sicily, Giuseppe was the only child left. The economic pressure on the family had eased. It was decided that Giuseppe, the baby of the family, should have the advantages his brothers didn't have to continue his schooling. The state was now paying for school for children until they were eighteen. Giuseppe at first considered accounting so that he could go to a trade school in his home town. Instead, he went to the trade school in Cefalù, a town near Palermo about thirty miles from his home town, to become a TV and radio repairman.

Giuseppa Maria and Bartelo Cascio, the parents, must have felt strange with all their children grown and gone. Soon they began to discuss moving. Gandolfo and Matteo feel that it was their mother who was the lonelier. Giuseppe, however, remembers his father talking about the loneliness he felt on Santissimo, Crocifisso, a big feast day.

"See," Giuseppe says, "in our home town my family belonged to a small congregation of eighty or ninety families. Each year the feast day is handled by a member of the congregation. It was my father's turn and none of his children were there to help them celebrate. After that, I think my father decided he was ready to move."

Should they try to move to Germany, where their three sons were, or to America, where their daughter and her husband were? The decision was to try to reunite the whole family in America. The elder Cascios made plans to move to Lawrence, Massachusetts where their daughter was living. In Lawrence their son-in-law had relatives and there were *paesani*–people from their home town. It would make things easier. The parents moved in 1966 and were soon followed by Giuseppe, who had just completed his technical courses. Next came Dominick in October 1967, and finally Matteo and Gandolfo arrived on December 23, 1967. Matteo and Gandolfo hadn't saved much during the six years they worked in Germany. They worked right up until the plant closed on December 22 so they wouldn't lose the annual bonus.

Matteo says, "We could have saved a lot, we earned enough in Germany, but we never did. How were we to know what was going to happen?"

Gandolfo says, "What are you going to do? We were young. We almost never ate at home, and we sent out our laundry." Pointing to his feet, he says, "We even sent out our socks. So we didn't have much saved."

"We were young then," Matteo adds, "and we wanted to go out and have a good time. The girls found us attractive. We always tried to be clean. Maybe they liked us because even though our clothes were not the most expensive, they were always clean. Our family taught us even if your clothes are not the most expensive, keep them clean."

"I wasn't that anxious to go," says Gandolfo. "I liked it in Germany, but my mother, it was very important to her. She wanted to move to America, so we got ourselves ready to go to America.

"As soon as Mercedes-Benz closed for the Christmas holiday, we came to America. Our brother-in-law took us the next day to get social security cards, and alien registration papers.

"My brother-in-law," Gandolfo says, "he worked in a clothing factory, but he suggested to us that we could make a little more if we went to work in the shoeshops. On January 2, 1968, seven days after we landed, we began working in the shoeshops.

"We worked in the shops for five years. People at Western Electric worry about the heat, but they don't know what heat is like if they haven't worked in the shoeshops. It gets to 110 and hotter near the furnaces. But we made good money for a while. It was piecework, and we worked hard, and we each made $10,000 or $11,000 the first three years."

"The best I ever did was make $301 one week," Matteo says. "I worked seventy hours that week, and I worked like an animal, but I made the money. Heh," he says, "I don't like to work, who does, but I like money so I work. I mean, what else can I do? I don't want to be a crook and make a lot of money the way lots of people do. For people like me to make money they've got to work. And when I do a job, no one does it better, but like it, no I don't like it."

The economic situation at the shoeshops began to deteriorate, and soon the $10,000 that Gandolfo and Matteo made was reduced to $7,000 or $8,000. The brothers then decided to look elsewhere for work. Both played soccer with local teams and both knew Nunzio DiMarco of Western Electric. Nunzio suggested that the brothers apply to work at Western.

In January of 1973 both Giuseppe and Matteo started work at Western Electric. Soon Gandolfo was hired and three of the four brothers were again working together. The fourth brother, Dominick, had a job as a barber in a shop in nearby North Andover.

When they first came to America, all four sons moved in with their parents. Gandolfo was the first to move out. He met Nina Aiello one and a half years after he had arrived in Lawrence.

"I met her at a banquet," Gandolfo says. "We danced and I said a few words to her. We started going out. Her parents and my parents knew each other. Seven months after we began going out we got married, and I moved with my wife into an apartment above my in-laws."

Matteo was the next to leave the family home. After Gandolfo began going out with Nina, both families saw more of each other. Soon Matteo was going out with Frances, Nina's sister. Seven months

FIRST ROW: Matthew Cascio. SECOND ROW FROM LEFT: Joe and Josie Aiello, Lisa Cascio, Giuseppa Maria and Bartelo Cascio. THIRD ROW FROM LEFT: Gandolfo and Nina Cascio, Irene Aiello, Frances and Matteo Cascio, Joe Cascio.

after his brother got married, Matteo and Frances were married. One year and five days later their first child, Lisa, was born. A year later Matthew—Matteo smilingly says, "It's the Americanization of Matteo"—was born. They now live in an apartment a few blocks from the Aiellos.

The other two sons remain at home. It is only in the last year or so that Gandolfo, Matteo, and Giuseppe have felt comfortable enough with this new country to think seriously about their futures. Each of the three has ambitions and dreams.

Giuseppe (Joe)

As soon as he came to this country, Giuseppe, nicknamed Pino at home, began calling himself Joe. For Joe life in Sicily, especially while he went to trade school, was very exciting. "Cefalù," he says, "was a beautiful place and there were young people from many counties in schools in the town. Heh, I really liked it a lot. I was being offered good jobs by companies, but I couldn't take them because my family wanted to move to America.

"I didn't know what to expect but I went. It was very hard for me. When you go from one country to another you don't know the language, you are like a baby. If I had been the first one here, I think I would have gone back in three months. It was easier for my brothers because they had already been in Germany.

"I thought, what was I going to do when I got here. I didn't think about school. I was nineteen, and then you think you are a man, you think you should work and make money. So I went into the shops, first the shoeshops and then the clothing factory.

"Slowly I began to pick up the language. Maybe I don't speak English so good, but I try. You may laugh at what I say, but only by trying can I learn. After a few years I began thinking about my future. I was no longer a kid and I started wondering why I was staying in the shoeshops. I don't have anything against people who work in the shoe and clothing shops, but I think maybe I can do better. So I moved to Western Electric. It's a big company with a lot of opportunities.

"When I began at Western I was already thinking about school. Sometimes I wish I were younger again, and when I first came here I'd gone to school, but what are you going to do? I decided to start school again. I didn't ask nothing of nobody. I just started taking a course last summer at Lowell Tech. Somebody told me that the company would pay for it. I found out they pay half the tuition at the start of the course, and then if you pass they pay the other half. So far I've taken four courses and got two A's and two B's. I'm going as far as I can. Right now I want to get an associate's degree in electrical engineering. I take three courses each term, and one during the summer. At my present pace it will take me three years. That's what I'm planning, but you never know what's going to happen.

"I'd like to get the degree, maybe even go for a BS. I figure by the time I get my associate's degree I'll have worked at Western for four or five years. People get to know you, you know them. I think they'll see I'm a good worker and help me out. In that way I think Western is a good place to work.

"Right now I'm living at home. You ask me how that is. It's fine. My parents treat me good. They don't interfere with my life. I don't think about moving out, why should I? I have nothing to prove. Some people in this country live away from their family because they need to prove something. But I lived away between fifteen and eighteen, and I feel I have nothing to prove. My parents say their house is my home until I marry, and that they enjoy having me. They won't let me contribute. If I brought something home they'd get mad. See, that is how it is in the old country.

"You must understand we live in America, but we keep some of the old values, especially my parents. My parents don't speak English, they speak Italian. Their friends speak Italian and we speak Italian to them. When they go shopping they go to the Italian stores. For them there is not so much difference here from their lives in Italy. That is the way it is for old people. But for us, the children, things are different. We must decide what type of life we want in this country, and whether or not we want to stay.

"I think that is why my sister and her husband moved back to Italy. My brother-in-law came here so they could make more money, but they always wanted to move back to Italy. If they think they can live better there than over here, then why not. Back in Italy my brother-in-law is working on a farm. When they lived here my sister worked. She didn't mind it because that is how things are done in this country. I don't know what to call it, but it isn't like that in Italy. In Italy the woman stays home. She takes care of the house and the children. That is how my sister wants it. So maybe they'll have less, but they will live the way they want. For them to move back it made sense. It may be hard on their children but they are still young and they can learn to adjust.

"For me, I'll make a life here. I'm in no hurry to get married. I'm still young. I have my work, and I have school. Already I attend classes three nights a week for three hours. I have a busy life. I play soccer and I go out. If I meet a girl and like her, how do I know I'll like her more than a girl I'll meet next week?"

Matteo

Matteo, like Joe, wants to get ahead. But Joe is only twenty-four, single, and already has a high school degree. Matteo is thirty-one, married, and has two children.

When Matteo first came to America, he didn't have the luxury of thinking about his future. He says, "When we came to America I had learned German pretty well. It was harder for me to learn English. I was older when I came here, and it isn't so easy when you're older to learn a language.

"The work at Western is pretty good. It is not so hard as the shoeshops. The pay is not great, but it isn't bad, and I can work some overtime. Still, I'd like to get a better job, a better position, who wouldn't? But if you don't have a high school diploma or any education, it's hard to get a better position, it's hard to move up to something like a technician.

"I hadn't worked at Western more than six months and one of the supervisors started talking to me about whether I was interested in a better job, in his shop. Sure I was interested, what do you think, but I'd like to be more certain of my English.

"So in July I started going to classes in English. I went two days a week, two hours a day at night. If I pass the exam I'll talk to him about the job. I don't know if I'll take any further courses, but I hope I'll have some more opportunities, and that I'll speak better English.

"Sometimes I worry about rising prices, who doesn't? For Italians, when you get married you have to provide for your family. That means that I'm supposed to do the work and my wife takes care of the children. I know that things are changing in America, but in my house I'm still the boss. I respect my wife, I listen to her opinion, but I make the final decisions.

"I prefer that she stays home with the kids. I think she feels this way too. Economically things are tough, but we are doing OK. When you get married it is fine if the wife wants to work, but once the kids are in the house I want her to stay at home with them. If we can live on my pay, then we will work it out. If we can't, then what are you going to do? If she has to go to work, then she has to, but then I think we'll have to work different shifts so there is at least one parent home all the time.

"See, kids are very important. When you have children you have to teach them. You can't just tell them things, what's right and what's wrong. You have to teach them by having self-respect and living a good life. I see plenty in the plant, old men chasing young women, but that's no good. That is why I think Gandolfo and I were lucky. As young men we lived away from home in a foreign land. That's why I don't need to fool around now. If I wanted to I could have lots of women, but I'm a married man, and you lead a different kind of life. When I was young and single I did as I pleased. That is why I think all young people should have experience like my brother and me.

"My brother Gandolfo and I are very close, more than a brother and a brother, more like a husband and wife. For thirteen years we worked in the same department and for the same company. For many years we lived on our own together. We think alike on many things."

Gandolfo

Of the three brothers it is Gandolfo I know best. He came to work a week after I did. When we were introduced, he was not called Gandolfo, but Andy. I finally asked him why everyone at work called him Andy. "When I came to work," he said, "the supervisor had trouble pronouncing Gandolfo, so I told him call me Andy. After that everyone called me Andy. What does Andy have to do with Gandolfo? Nothing. You know," he continued, "in that way Americans are a little bit spoiled. If I were somewhere, even in China, I'd learn how to pronounce someone's name.

"When I came to this country I was twenty-seven years old. I wasn't a young man anymore, and already I'd worked for thirteen years. First thing we did when we came was to find a job, then we could worry about what we wanted to do. It took me longer to learn English than German. I still don't speak English good. By the time I began to learn things, I got married and had responsibility, a wife. Sometimes I think about if I could write better I would get a better job. I'm not complaining, but it is too late for me. I missed something, but I'll send my children to school.

"That is, my wife and I would like to have children but not so many. I'm not the kind of Italian who wants the big family. We don't need four or five children. The church may not like it, but you have to make your life, you know what I mean. If you have a big family like that, it is just like the old country; your children don't have a chance for school. But we would like to have one or two children. So far we haven't had any luck.

"Right now my wife works. Since we don't have children, I let her do what she wants. What do you think, if I told her to stay home," he says laughingly, "what do you think she's going to do? She likes her job. She works as a secretary just a few hundred yards from the house. She is never tired when she comes back from work, and that makes me happy.

"I'm less tired now that I work at Western. Even on a hot day it is nothing like coming home from the shoeshops. I may make less money now at Western, but the work is better. With my wife taking home about eighty dollars and me about a hundred we don't need much more. Sometimes now I work overtime on Saturday morning. My wife goes to the beauty parlor on Saturday morning. What am I going to do, lay in bed? So I go to work and I'm home by twelve-thirty.

"My job at Western is all right. They offered me an upgrade. They wanted to know if I wanted to bid for a 33 grade job. I told them no. I would have to go on the second shift, and to the other plant in Lawrence. I didn't want to work the second shift. If I wanted to work the second shift I could have started work at Western nearly a year ago. Maybe I'd have taken the job if we had children, but we don't need the money.

"My ideas have changed a little. When I was young I was a big strong kid, full of energy. I never used to think about retirement, now I do. Not that I am old, but when I get to be sixty or so and if I have a chance, I'll be ready to retire. I tell my mother-in-law that, I tell her she should quit and just relax and have a good time, instead of keeping on working. I'll find plenty to do when it's time for me to quit."

It is difficult for Gandolfo to talk about himself. It was from Joe, and especially from Matteo, that I learned how hard he worked when he was young to help the family, how he was a leader of a strike, how he was a top worker on the German construction site. Gandolfo does not talk about how hard he works. He says he doesn't work hard, that he takes breaks like everyone else. Yet everyone notices what a good worker he is.

Almost the only time you can get Gandolfo to talk about himself is when he talks about soccer. One day he came into work especially happy and told me that he had scored a goal. "You wouldn't believe it," he said, "from a corner kick, Yanick put the ball in, and then," snapping his head to show what he had done, he said, "I headed it in. I think it's the best goal of the season. Just like Pele did in the world championship."

To appreciate what Gandolfo is like, you must see him on the soccer field. He moves around the field with precision more than grace. There is not much about his play that can be called exciting. Unlike his brother Joe, voted the league's outstanding offensive player, there is nothing flashy about Gandolfo. He is more like a turbine than a gazelle. He always seems to be in the right place, always moving, never tiring. Up and down the field he goes, passing, setting up, coming back to protect the goalie.

On the night of the banquet he stands with his two brothers in front of his friends from Western Electric. Three times he is called up for trophies: once for being the league's best defensive player, once for being the league's best referee, and, most importantly, for being the team's choice as best player. His basic soccer skills are certainly no greater than those of his teammate Yanick, but there is something about him that is compelling. There is a strength he had clearly carried for a long time, a strength from his Sicilian childhood.

5. FATHER AND SON

At seventeen MIKE BEAL was a squad leader in the Army, giving orders. At twenty, when I met him, he was a 32 grade bench hand, taking orders. Mike and I began work together. Until we found groups within work that accepted us, we had each other to depend on and to relax with.

Willard (Bill) Beal is forty-six years old; his son Michael is twenty. Bill has worked for Western Electric for twenty-two years, his son Michael recently worked for Western Electric for nine months before quitting. Bill is settled. He holds two jobs and has helped raise five children. Michael, just out of the Army, has barely begun his work life and is unsure about what he wants to do.

"It was 1951 when I started at Western," Bill Beal says, "the day before Christmas. I worked a half day and got the holiday. It was only my second job around these parts here. I was working at Hytrone and I took a day off and went down and applied to Western. They wanted me to start right then and there, but I told them I had to be fair to the other company and give them two weeks' notice, which I did.

"I started at the old Haverhill shops on one of the drill presses. I did well and I went from 32 [grade] to 33, got 34 and went over on the screw machines and then I made 35. Then they started building the North Andover plant. There were two men ahead of me in seniority and they offered them a setup man's job. Both turned it down. They were getting a little older and probably couldn't handle the job. I got the opportunity and I got a double jump in grades. I went from 35 to 37.

"Then I was a 37 grade for quite a few years and then we started doing engineering work. We had to do a lot of figuring and making a lot of form tools and that's when I started to squawk. I thought I should have a 38 grade and the union backed me up on it. They negotiated it and I got the 38 grade after promising not to bump the guy on days.

"I worked on the evening shift for fifteen years before moving over to days. The shift really doesn't make that much difference to me. You go into that place and it's so big, you don't know what it's doing out, if it's raining, snowing, sleeting, or if the sun is shining. I like what I'm

doing anyway, so I go in there and get busy. I don't worry about what the weather is outside.

"I'll tell you, I like my work. I wouldn't say the majority of people I work with like what they're doing, but I do. I look at things a little differently than most people. I'm not saying that I owe the company, because I don't. I did the work and I did a good day's work. Ever since I've worked at Western Electric, I tried to get good things. I got two cars. I started off in a new car right off the bat. I've raised a good family; I got five kids. I built a camp. I've owned two homes. So I really can't complain because I have made a good living. That's what everybody works for, right? To more or less survive and make a good living, which I have.

"It hasn't always been easy. I think Mike told you I work a second job. I'm not complaining, as I said, I like good things and the extra money comes in handy. I work for a guy who used to work here at Western. I work three or four extra hours a day, four or five days a week. It's a good job, and the guy I work for, he was the one who asked me to work for him. I didn't go to him to look for a job. He called me up one day; he had two screw machines in and he didn't have anybody to set them up or run them or anything. So I told him I'd help him out.

"But work has never bothered me. If I'm not working to get paid, I'm always doing something around the house. I like to keep myself occupied. That way if I feel like taking a nap on Saturday afternoon or a Sunday afternoon, I take it. I think I've more or less earned it.

"So, all in all, I don't have any gripes. A lot of guys, it takes an awful lot to satisfy them. I'll admit I'm doing something I like. I don't like it all the time, but I like my work. I can understand a lot of people not liking what they're doing, especially bench workers.

"I don't care what anybody says, even when I get on the bench, it does get boring. I don't care, bench work to me is so goddamn boring, it's pitiful, and I think a woman can adapt to it more. Maybe it's part of their personality, maybe it's built in them, I don't know, but a woman don't seem to be bored at all. Some are, but I'd say 99 per cent of them aren't. I never heard of any women saying, 'Gee, this is a boresome job.' All the young fellows over there right now, that's doing bench work, they're bored stiff. I think that's what happened to Michael, he couldn't stand it, and that's why he quit.

"I wouldn't want to hold him down on an inside job if he's not going to be happy. He talks about working outdoors, something to do with forests or the woods, animals. That's what he likes. I understand his point of view, but I also want him to look at my point of view. He had a good job in here, it was a steady job, which is important. There are good benefits in here. In fact I had my insurance agent at Metropolitan look into it and he claims that this is one of the best benefit plans anywhere in New England. They're that good and that's important. There's a lot of security working here, which you don't have on many jobs.

But Mike says he's not worried about those things now. As he told me and his mother, 'When I get married and I have to start worrying about supporting a family, then that's when I'll start thinking about benefits and security.' Right now he says he wants to do something he likes.

"I can understand that, but I don't want him to think you can have a job you always enjoy, especially not at the start. I had to work on a job I didn't like for two years.

"I guess I was more ready than Michael to settle down when I started at Western Electric. I was born around here, over in Bradford, and I went to Haverhill trade school. There were a lot of things I wanted to do, so I quit school and went into the Navy. I think everybody should go into the service. It helps them grow up, in a lot of ways.

"Now when Mike went into the service he was more on the kiddish side; when he got out of the service, he was more on the mannish side. I can tell by his friends that haven't been in the service because there's quite a difference in the way they act, the way they do things, the way they think. I think it does a boy good. It helps him grow up and understand his life. I think they get a lot out of it that way.

"When I got out of the service I had the money I saved and I took off around the country. I hitchhiked out to South Dakota, and a buddy of mine from the service and I took jobs in the gold mines. The idea of working in a gold mine fascinated me right off the bat. I started as what they call a mucker, pick and shovel, laying track. Then I worked my way up to what they call a loader, and was operating these compressed air shovels. I worked there for one and a half years and I had a chance to see bottom; it was five thousand feet underground. When I had my bellyful of that I just pulled up stakes and left.

"We went to a ranch next. We worked for my friend's uncle, who owned a big ranch in Alzeda, Montana, for three months. Then we headed to the corn belt, Akron, Iowa. We worked up on a farm up there for about six months. I learned to milk a cow. I learned how to get up early in the morning and do a lot of chores before you sat down and had breakfast. I found out what it was like for a farmer and how he had to work and survive.

"When we left there we traveled. I saw lots of things like Rushmore Monument, the Needles, Custer's Last Stand, Deadwood, South Dakota. I kept traveling for a while, and would have done more but my mother got sick and I had to come home. While I was home I met Carmie, and we got married, but not till I worked another six months to get a foothold, and then, well, you know I started at Western Electric.

"I think it would do Mike a lot of good to go out and do what I did, meet other people, and see how they feel about different things and see how they live. I don't mean the rich people. I mean people like myself, the lower class, you might say, or middle class, whatever you want to call it. And see what they have to do to get by and survive. To me, that's a good lesson. I've worked with guys who have never done that. They take too damn much for granted. I think that's what Mike does right now, he takes a lot for granted. I think that's what's his problem. I think he should get out more. He seems to think people owe him something.

"I don't know, something is eating at him and it's tearing us apart. I know I said the service is good for a kid, but Mike and me, we've been having problems ever since he got out of the service, and it's breaking my heart.

"I've always thought we've been very close together. Anything that would happen to Mike I was really interested in, and I was usually there when he was in trouble. We always done things together like build the summer camp. He helped me with the whole thing. And we've taken trips together, just the two of us, like we took seven days and went up to Maine, we took off and did a lot of fishing. We put over seven hundred miles on the car in a week. And no matter what I was doing, we always tried to keep Saturday morning as our morning. We'd go out and get a breakfast, wheat cakes and stuff. We always looked forward to our mornings together. But they grow up and they forget a lot.

"Things have just changed since he got home from the service. Maybe some of it's my fault. When he got out I had the tendency to keep him home. I wanted to have him help me do this and help me do that, without even considering how he felt about things. But a lot of it is him, he's always ready for a fight. Since he's gotten out of the service we argue all the time. I told him, you're not in the Army when you're in this house and he said, 'You always say that.' He gives his mother a lot of back talk, which he's never done before. This is stuff I just don't understand and there's no reason for it. He can eat three good meals a day here. He's got a good place to sleep, and I think he's got a pretty nice home. His sisters are good to him, we all love him but he has an odd way of showing he loves us.

"It doesn't seem to matter what you say, he gets mad. You know, he's all wrapped up in his band now, he plays the harmonica. He's never taken lessons but he's damn good. For a while he was practicing in the house. I had to stop him. I didn't mind his playing but they play loud and I can't have it around. When I work twelve hours a day, I can't come home and listen to it. I'm in bed by eight-thirty or nine o'clock and I've got to get up bright and early, so I like a little quiet. Well, he was mad.

"But things keep getting worse. Things have gotten a lot worse since this summer and an incident up at the camp. He started picking on his sister at the dinner table. I told him to shut up. I said, 'You do the same things.' He said, 'I don't have to listen to you.' So he got up and left, walked right out on us. Which I don't think there was any call for because he was way out of line to start with. Because he does the same things that he complains about. He floored me when he did that. I told my wife, he broke my heart when he did that, and he's not going to do it again. I'm not going to take the chance.

"Talking to him right now is like talking to a wall, until he snaps out of it. So I don't talk. He can show me a little respect anyway. He always did; there's no reason why he should change. Because I've always treated him the way he should be treated. I gave him the benefit of the doubt every time. I don't understand it. I don't know what to do. I don't want Mike to forget us altogether. A close family is very important to me and my wife. Her family's always been close. Mine wasn't that close, but I've tried with the kids, and with my brother. My parents got divorced when I was in the service. I still can't understand that. They were married for twenty years and then all of a sudden it's gone. They bring you up a certain way, they teach you a certain way, and then all of a sudden it's gone.

"I like our family being close, and Michael's an important part of it. But like I told you, this is my house, he has the option to go and live on his own if he wants. I won't hold him. But as long as he's under my roof, he has to abide by my rules. There are always going to be rules regardless of where he goes.

"I just don't seem to reach him. Really, I think he's just bored. He's bored with a lot of stuff. There is no action for him around here. He's said that a couple of times, there's just not enough action for him."

Michael Beal was just out of the service. His father had helped him get his job at Western. The first few weeks Mike and his father had lunch together almost every day. Mike talked a lot about his father. He was worried about how hard he was working, holding down two jobs.

"You know," Mike said, "before I went in the service my father could do just about anything. But he's really kind of tired these days. Working two jobs takes a lot out of him. He doesn't have as much energy. I tell him that he should stop the second job, but he won't listen. He tells me we need the money for the camp, and the boat. I tell him what good is the money, he's going to run himself down and he won't be able to enjoy it."

During a smoking break, one afternoon, Mike introduced me to his father. Bill mentioned that he had four children. I casually remarked that I hoped the others were better than Mike. He took my joking remark seriously and, putting his arm on Mike's shoulder, said, "I'll be glad if they turn out as well as Mike."

Mike worked very hard the first week, harder than I did. But after a month had passed his attitude toward the job had soured, and his disposition with it. He found the work we were asked to do boring and monotonous. One day he told me that he burned his finger because he wasn't paying any attention to his work. He kept saying he couldn't stand being cooped up, that he wasn't suited for the work. As the weeks of the summer went by, he became more and more short-tempered, more anxious to get out.

One day in sheer frustration, he asked Tom, an older worker, "Do you think I should take the 33 grade coil wrapper job in Lawrence?"

Tom said, "No."

Then he asked, "How do you think I can get ahead?"

"Quit, go to school."

"I can't, I need the money."

Bill Beal, December 1945.

Mike in uniform.

Tom shrugged his shoulders and said, as he walked away, "Well, then you'll just have to keep on doing what you're doing now."

Mike was still working when I left Western, but he was complaining about it more often. Three months later, he quit and took a job delivering milk. He held the new job for only four months. His temper got the best of him one day at work and caused him to be fired.

Michael is unemployed now, but is thinking about going back to Western for another try. He doesn't see how he can work things out on the job until he clears up his personal relationship with his father.

"My father and I were real close," Mike says, "but not any more. It really changed since I got out of the service. See, I didn't come back the way he thought I would. He wanted me to come home like Joe. Man. You know, I'm still twenty years old. There's a helluva lot that I don't know. The Army doesn't teach you everything about life. He seemed to think I was going to come back like him; just settle right down, grab myself a woman, marry her, have a few kids, buy a house, the whole thing. But I'm not like that. I want to be free for a while.

"I'll tell you one of the big reasons I went into the Army was because I thought my father would like it. He was always talking about the Navy, so I figured I'd go into the Army when I got out of high school. It made him happy, I guess, but he never really told me exactly what he thought about it.

"So I joined the reserves. I was only seventeen years old. They sent me to Fort Jackson, South Carolina, for basic training. They made me a squad leader; that meant a lot to me. I was giving orders to guys older than me. From basic I went to A.I.T.—Advanced Infantry Training. I became a squad leader down there too. I graduated fifteenth in a class of almost seventy, and decided to extend my reserve active duty.

"I was sent back to my basic training area, as an assistant drill sergeant to Drill Sergeant Rhodes (SFC), the D.I. who put me through basic. I wanted him to see how well I had turned out.

"One day I went up to Sergeant Rhodes' apartment. He had pictures on the wall, right, like his wife and his daughters, and then there was one army picture on the wall, and it was a platoon picture. He had put hundreds of cycles through but there was this one picture, the platoon I was in, and I was the first squad leader. It blew my mind; we must have been a good platoon for him to want to put our picture on the wall.

"I don't want to brag, but a lot of drill sergeants down there wanted me to stay on active duty as a D.I. I thought about it, but finally I said no because I was homesick for familiar grounds, my friends, and my family. I did eighteen months and then came home.

"When I came out of the service I found my father was lecturing me all the time. I guess I never really noticed it before but being away made it really noticeable. When I came back from the Army, after being the man for a while, I realized my father is pushing me around. Maybe he doesn't realize it, but that's the way he acts toward me. I do something in the house, he says, 'Don't do that again.' For Christ's sake, I'm twenty years old, I've been in the Army. I was the man. I was a drill sergeant, a leader and he still lectures me as if I were ten years old.

"My father is a terrific guy. I mean everyone likes him, but we can't get along. Things are getting worse and worse between us. He forgets I'm an individual and need to decide things for myself. I don't think he even knows but he's holding on too tight. He's holding on so tight that I've got to yell, 'Let go of me.' He's squeezing me so tight I have to get out. If I don't I'll get crushed. Me and my mother talked the whole thing over and we came to the same conclusion, that I should leave home. She doesn't want to throw me out but she says it would be best. It's tearing her apart seeing me and my father arguing all the time.

"Things started getting real bad this past summer at the cabin. It was one of those stupid little things, one of my sisters was eating too loud. I told her to stop eating like a cow. I guess I said a couple of other things. Anyway my father started lecturing me about my table manners.

"He pissed me off and I told him, 'I don't have to take this shit,' and I left for home. Later he showed up and we started again. This time he hit me, something he hasn't done in years. I wanted to hit him back, but he's my father. Instead I smashed a high chair on the kitchen floor and stormed out of the house.

"I knew then I had to get out, and work these feelings out because they were getting in my way. When I quit Western I didn't tell my father because I knew he wouldn't like it. He would have started on me, how it was such a steady job, and how I should stick it out. I didn't want to hear all that. I hated the job. I was sitting on my ass, putting on weight. I couldn't stand being inside, and the job was so boring and monotonous. I wasn't going anywhere. I would've been stuck on that bench for a long time. It was driving me crazy.

"He says he understands that, he knows how boring bench work is, but still I know he'd want me to stay. He'd tell me you can't always have a job you like. He'd tell me how it was boring for him when he first started. Maybe it was boring for him when he started, but he's doing what he likes now. He likes working on those machines. He's got friends at work. He goes into work, stands around tending those machines, goes into the men's room and has a couple of smokes. He's got a little pull at Western, so it isn't so hard for him.

"I didn't want him lecturing me, so I quit and didn't tell him. I guess I was afraid of disappointing him. I've always tried pleasing him, and it gets in the way. When I was in high school I wanted to wear my hair long but I didn't because I knew my father wouldn't like it. I got a job working six hours a day after school not only for the money but because I thought my old man would think more of me.

"This whole thing has given me a wicked temper. My temper got to me while I was delivering milk and got me fired. I'm beginning to wonder if maybe I should see a psychiatrist, because I know I've got a problem, an authority type thing. I can't stand anyone telling me what I should do, or ordering me around. It's not actually the being told, it's the way it's done. If someone tells me the way my father does, that's when I fly off the handle. I've got to learn to control my temper. I've got to start making my own life and my own decisions. If I want to grow my hair, or play in a band, I'm going to do it because I want to, no matter what my father thinks. I don't know, maybe we can get back together and be close the way we were, but it's got to be man to man."

Running
Somewhere
?

DON'T
RUN!

Western Electric, with nearly ten thousand employees at Merrimack Valley, believes it could not operate with any sort of efficiency without establishing rules. Along with pamphlets on medical coverage, life insurance, and tuition payment plans, a new worker is given a pamphlet on rules. The pamphlet begins with these words: "When a great number of people are working together certain rules and regulations are necessary so that the business of the Company can be conducted in an orderly manner and the welfare of employees protected. These rules are founded on common sense and the principles of good citizenship."

While working at Western Electric I learned a great deal about the rules and workers' reactions to them. Probably no one thing affects and bothers shopworkers more than the time clock. Each person in the plant is issued a time card that is to be punched in at starting time, before and after lunch, and at the end of the day.

The first time I saw someone punch in late, the girl next to me said, "You know, they won't pay you for punching in early, and they won't pay you if you decide to stay an extra ten minutes late to finish a board, but just do like Jan did, punch in a minute late, and you'll see that minute missing in your paycheck." Repeated lateness can cost you more than money from your paycheck. It can lead to a union notice, and it is sufficient reason to be terminated.

I remember the first time I was late. About two months after beginning work, my alarm didn't go off one morning. When I woke up it was 6:00 instead of 5:15. I panicked, let out one "Oh, shit," hurried through brushing my teeth and slipping on my clothes, and was out the door in less than five minutes. I grabbed some bread and ate it on the way. While pushing the car over the speed limit, I kept an eye on the rear-view mirror for a policeman, my heart pounding. It usually takes me thirty minutes to drive to the plant, so I was almost certain I'd be late, but I drove like a crazy man, rolled into the parking lot at 6:27, and raced into the plant. Since you aren't allowed to run, I was reduced to a laughable form of quick walking. Somewhat out of breath, I reached for my time card only to discover that someone had punched me in. Later in the day a woman told me she had done it. She said, "I figured you'd get here. The company won't miss the money and you can watch out for me."

The next time I got up late, I was not as fortunate. Though I rushed like mad again, I was about five minutes late this time. I clocked in and went to my seat, relieved that I had gotten in so close to the bell. Jack[1] came by my bench and said, "Glad you could make it in today."

I related my anger at our supervisor's remark to an older worker. He shook his head in disbelief and said, "Look, they aren't going to pay you for rushing. If you're late, you're late. I figure the couple of times I'm late, I'm not going to kill myself. Hell, I've been here eighteen years and they aren't giving out any medals. I just take my time, make my breakfast, and come in late."

Just as punching in late is against the rules, so is leaving too early. People begin cleaning up about five minutes to three. Their work places cleaned up, their tools put away, they edge toward the back of the room to the time clock. By 2:59 a majority of the people are lined up at the clock, their cards poised in their hands, waiting for the three-o'clock bell. As it rings, cards fly into the time clock, are placed in their slots, and people are off on a mad dash to their waiting cars.

Many people have resorted to running in an attempt to beat the rush to the crowded parking lot and the unavoidable delay. The running has led to occasional accidents, and the company, which has long had a rule against running in the plant, decided to enforce it more strictly.

The company began its crackdown on running by sending a memorandum around restating the rule. It then sent photographers around to document running. More than once I had to assure people leaving the plant that I wasn't a company photographer. The company finally posted supervisors along the corridor to act as monitors.

The new enforcement stopped people from running in the plant. However, a variety of quick walking techniques have developed. With the supervisors standing in the corridors, people slow down until they pass the guard stations. Once people are in the parking lot, their feet start moving more and more quickly and the walk soon turns into a run.

The only other way to avoid the rush is to leave early. If you leave early and clock out properly, you lose money, so occasionally a worker will leave a few minutes early and ask another worker to clock him out. I remember one woman in our area nervously waiting for the three-o'clock bell. She wanted to get home to see her new grandchild and didn't want to fight the crowd, so she asked me if I would punch her out. I said, "Sure," and five minutes before three she began walking to the exit near her car.

[1] The names in this chapter have been changed.

GATE No. 6

The bell that rings to start the day, and to end the day, also rings to mark the beginning and the end of the day's two ten-minute breaks. You are supposed to be in your seat by the end of the break. Hypothetically, this means you'd have to begin to move toward your chair before the bell announcing the end of the break rings. In practice, people don't begin to move back to their work areas until the bell has rung. Some supervisors are very strict about starting work as soon as the break is over, others aren't. Jack's reaction depended upon how production was going and his mood. Generally, he didn't bother us as long as we returned to our work area promptly when the bell rang. However, if you were a little late, you might sometimes find him hovering over your bench waiting for you. He rarely said anything at those times. He just looked at you with stern disapproval. I was always astonished by how effective Jack's looks were at making adults respond as if they were errant children. I often found myself walking more quickly if I saw Jack near my desk at the end of a break. Like most other people, I would hurry to sit down and to begin work, although not a word had passed between us.

A worker is required to punch out and back at lunchtime as well. Although punching someone else's card in the morning or at the end of the day did not occur often, it was done with some frequency around lunchtime. With only thirty minutes for lunch, the people who ate in groups in the shops would punch each other out. One person might leave a couple of minutes early to get drinks at the vending machine or from the cafeteria. When the bell rang, a friend would punch his card out.

We were also required to punch back in before the half hour was up. So people often wandered back to the shop five or ten minutes before the end of the lunch break, punched back in, and stood around smoking and talking until the bell ending the break rang.

The time clock restricts personal movement by reducing the day to periods in which things are allowed and not allowed. It routinizes the day by regimenting personal needs to meet the workings of a bell.

The time clock undoubtedly curtails working time lost due to people coming in late and leaving early. Its presence and the rules that surround it, however, create a good deal of hostility among workers. As one woman told me, "In here we don't ask for whom the bell tolls, but for what. The bell tells us when to smoke, when to eat, when to start work, and when to stop." Another woman told me, "They've got the clock, and we're forced to live by the clock and their rules." One way many workers react to the ever-present time clock is to figure out

ways to "beat" the clock, to remove at least part of an occasional day from the clock's control.

The company sets aside forty minutes during the day for a person to take care of personal matters such as going to the bathroom or going to the vending machines. Twenty of these forty minutes are taken up by the two ten-minute breaks. So one is theoretically left with twenty minutes during the day to take care of personal matters.

A sense of how free you are to use that time or take additional time is greatly influenced by the supervisor's attitude. Several women told me that they had worked for bosses who made them afraid to get up. They said they almost had to ask permission to go to the bathroom. One day at lunch one of the fellows said, "There's a guy in our shop watching us all the time. He's supposed to be watching our production and figuring out why it's so low. That S.O.B. has got it so we have to even sign out when we're going to the crapper."

Our supervisor wasn't like that. We were pretty free to go to the bathroom or vending machines whenever we wanted. Nevertheless, most supervisors don't like to see people unoccupied, so if you are finished with a job, or restless, and can't find something to do to occupy your hands, the bathroom is the place to go.

Many people use the bathroom as a place to read a morning paper or, like me, a small book or magazine. Some people boldly carry their newspapers in a back pocket; others carry things more surreptitiously. For a week I watched one man slip his newspaper under his shirt each day before going to the bathroom. I knew a woman who put small magazines, movie magazines, in her purse. I carried my paperbacks in a back pocket.

The bathroom is also the place where an important but infrequently enforced rule is broken. Smoking is restricted to certain areas and to certain times of day. Basically you are allowed to smoke only during the two breaks and lunch. There are a lot of heavy smokers working in the plant, and the bathroom is the place where many break the no-smoking rule. The day I began work a new worker asked another worker, John, if there was anyplace to take an occasional smoke.

"I might as well tell you because you'll find out soon enough," John said. "Although it's not allowed, you can get away with it. Most of us smoke in the bathroom. It's hell," he said, "if you're a chain smoker, but if you're a once-an-hour smoker, you can do fine."

Therefore, the bathroom is often filled. I can't speak with any authority about the women's bathrooms, but in the men's bathroom, there are times during the day when you are in big trouble if you need a stall. You have to wait, listening to the rustling of paper, and watching smoke roll over the closed stall door.

The rule against smoking is bitterly resented by large numbers of people I talked with. There are other rules that are not as intensely unpopular but that, like the rule against smoking, the company has chosen to enforce selectively. Two rules that come to mind are those against taking collections and against selling things in the plant.

There is a rule that states that there are to be no collections made in the plant. The rationale for this is that if collections were allowed many people who otherwise might not give would feel some social pressure. [In a plant employing ten thousand people, it would be a rare day in which someone didn't have a family tragedy and a need for a collection.] Aside from the United Fund, which is aided by management in its effort to collect in the plant, plant-wide collections are restricted to very rare emergencies.

Yet within the small work groups, collections are permitted. There are collections for going-away parties and for serious illnesses or deaths in a worker's family. Often cards acknowledging the collections are posted in the work area.

Just as the rule against collections is selectively enforced, so is the rule against selling things in the plant. Again, the existence of the rule makes sense. If there weren't a rule, there would be a thousand and one things sold in the plant, from toiletries to insurance. People are aware of the rule, but they break it. In the area where I worked, there were two small businesses.

One woman sold ties which her sister hand made. They were very popular and sold for $3.50. On the front of the tie you could have a name inscribed in reflection, creating a nice pattern. On the back, you could have a hand-painted picture of your spouse, or girl friend, in a bikini or in the buff. It was a popular item.

The other business was in used golf balls. One of the younger workers in our group went golf-ball hunting with a couple of his buddies. They would tour the water traps of golf courses around Massachusetts, Vermont, and New Hampshire late at night and retrieve errant balls.

In July, Joey began bringing in small bags of balls. He would sell the best for three for two dollars, others for as low as three for one dollar. The word spread quickly, and soon Joey was being visited by men from all over the shops. Such small businesses could be found throughout the plant.

There are some rules that the company feels strongly about and tries to enforce. These rules include drinking in the plant, taking drugs, and gambling. Being caught drinking or using drugs on company property can lead to immediate suspension. The company is aware that some of the company's employees do have drinking and drug problems. A counseling service has been established to deal with those who admit to the problem. Strong rules in these problem areas supposedly keep violations down.

I saw little evidence of people being high. In fact, I saw only two young men who were obviously high. One, I was told by another friend, was on speed. He looked like he was flying the few times I saw him. A couple of times I heard people talk about grass or uppers. In the john the talk was about getting "bombed" or "wasted" on booze over a busy weekend.

I never got any real sense of what the drinking and drug situation was or even if a real problem existed, so I kept trying to find out from other workers. I did discover that most people thought drinking was a far more serious problem in the plant than drugs.

Lots of workers had stories about fellow workers they knew who had drinking problems. There were some funny stories about people bringing liquor into the plant in their thermoses or of people who liked to take a stroll to their car after lunch for a "little nip."

A couple of workers had caustic remarks to make about supervisors and drinking. I was told to get a "whiff" of this or that department chief's breath after he had returned from a social luncheon held outside the plant. The way these remarks were made led me to believe that they were as much about long, leisurely lunches, a privilege that isn't extended to shop employees, as about drinking.

Being caught gambling can also lead to immediate dismissal. The kind of gambling the company is most concerned about is the numbers, betting on the ponies, and other racket-connected gambling. The company is aware that such a large, captive population is attractive to racketeers. The company claims it wants to protect workers from gambling above their limits.

That doesn't mean that no gambling goes on. For the most part, however, it is limited to small bets. For a period of three weeks, three of us had a dime bet at each break as to who could throw his cigarette butt into the receptacle. The winner took the pot. One woman who played with us was a consistent winner. After a series of amazing shots, she collected her pot and said, "Well, I've been here long enough to be good at this."

There's a lot of card playing during lunch. Supposedly money is not to change hands. It does, but not openly. The gambling restriction seems to keep the betting down.

Most workers I've talked to realize the need for rules. However, a number view some of the rules the company has, and the way they are executed, as oppressive. One day Jean, who usually seems very relaxed about the rules, told me, "There are too many rules here. It's just like being in school. There's one place to smoke, one place to do this, and one place to do that. If you get out of your seat to talk to someone, you're breaking a rule."

When someone was once reprimanded for breaking a rule, we got into a conversation about rules during the afternoon break. Mary said, "You know, things are changing; the kids won't stand for the stuff that we take. If Jack gives me a look I'm back in my seat immediately. But not Diane; he can look at her all he wants and she'll take her time. It

isn't that I want to go back to my seat, it's just that when I was brought up we were taught to follow what your boss or father said, and no questions. The kids now are different. They aren't afraid like we were."

Joanne said, "I'll tell you. The kids have made the company change. Ten years ago if you wore your pants a little too tight or your dress too short, they'd send you home. But the kids started squawking and now there really isn't a dress code."

Ruth looked at me and said, "Ten years ago they would have fired you for wearing long hair, but kids won't cut their hair for the company anymore." Joanne added, "I think it had to do with the Vietnam War. After the war things started changing in here."

The bell marking the end of the afternoon break sounded. As we headed back to our seats, Howie said to me, "They don't have any walls in some of my kid's classes. It's one of those experiments. They let the kids make their own schedules. You think my kid's going to put up with the kind of rules they have around here after that?"

Probably not. As the schools change, as the structure of the family changes, and as patterns of socialization change, so will factories. The company has already been affected by the demands of younger workers for looser structures. As those demands continue, more changes will probably be made.

Nevertheless, the company has set up an intricate set of rules designed to minimize distraction and maintain a high production schedule. In such a system there are bound to be tensions between the company's needs and the individual's needs.

By setting rules as it has, management has helped foster an atmosphere of *we* and *they*. The company is the all-powerful source of authority, and it becomes *they*. The workers required to live within the structure the company sets, become *we*. This duality is constantly felt and expresses itself in a variety of ways.

For example, a new worker is usually shown the "ropes" by other workers. This includes being told about rules and ways to break them. This shared, subversive information helps build a sense of solidarity among workers. The breaking of a rule represents more than the simple disregarding of an order. The action often serves to allow a worker to express some sense of personal control over his environment, as well as to demonstrate and promote solidarity with other workers.

One of the clearest examples of the multifaceted nature of rule breaking is the unpopular no-smoking rule. Many workers see no reason why they shouldn't be permitted to smoke at their work places, especially when office workers are allowed to smoke at their desks. Outsiders often suggest that the company should demonstrate concern for the workers by setting up reading and smoking areas.

Though that probably should be done, smoking in the bathroom has evolved into something more than simply breaking a rule. Somehow, having a smoke in the bathroom bestows a sense of safety, of being away from the supervisor's eye for a few minutes. The worker has found not only a place to smoke but, more importantly, a place that has not been designated by the company. Since a worker often feels that much if not all of what he does is done in places designated by the company, under company control, finding ways to express personal freedom from this institutional regimentation is important. When the day is so structured, being able to do something illegal can become very satisifying.

The inviolability of this area is generally tolerated. Once I was in the bathroom when a supervisor walked in; everyone felt ill at ease. The supervisor quickly left without saying anything about the smoking.

SMOKERS
PRACTICE
GOOD
HOUSEKEEPING
"PUT 'EM HERE"

UTION
MATCHES
SMOKING
IN THIS

Within certain limits, the breaking of rules is not very disruptive to the company. It is incredible, in fact, how infrequently major breaches of rules occur within the plant. If the company chose to enforce all of its rules strictly, workers would undoubtedly react in more negative ways toward their work. Sabotaging work, which I never saw or heard about, might become a big problem. People can be pushed only so far. To have restrictive rules that can be broken to a limited extent allows a necessary escape valve.

No large institution could function without rules. Many of the company's rules are widely accepted and seem to cause little resentment. Rules against drinking and drugs, against collections, and against selling, cause few problems. That is because they are rules that are not directly related to the work situation. The company is viewed as being on legitimate ground when it bans these activities from its premises. The tensions develop around those rules that are directly related to the work situation. A worker is constantly confronted by these rules, and it is inevitable that many of these rules cause deep resentment.

By establishing so many work-related rules, the company is tacitly admitting it believes that without such rules the work to be accomplished would not be done. It coaxes production along not only with a monetary reward system, but by severely limiting personal initiative and choice during the working day. It tells the worker not only what to do, but what not to do, and when not to do it.

It is impossible for the worker not to react to these restrictions and to be very sensitive to encroachments on individual freedom. No other segment of the working population has its day so closely regulated and watched as the blue collar worker. Perhaps this is one reason why blue collar work is ascribed relatively low status in our society.

A delicate balance exists between the company's need to keep order and maintain high production and the workers' needs not to feel powerless and impotent. The need to break rules will remain as long as workers exercise little control over their immediate environment. There are increasing demands to move away from the autocratic, top-down, rule structure that exists. These demands will continue to mount as younger workers who have had fewer autocratic institutional experiences move into the plant.

THIS MEMO, LIKE MANY OTHERS, WAS CIRCULATED BY EMPLOYEES AT WORK.

To: ALL PERSONNEL

Subject: ABSENTEEISM

It has been brought to our attention that the attendance record of this department is a disgrace to our company, who at your request has given you this job. Due to your lack of consideration for your jobs with this company, as shown by such frequent absenteeism, it has become necessary to revise some of our policies. The following changes are in effect as of today.

SICKNESS- ABSOLUTELY NO EXCUSE: We will no longer accept your doctor's statement as proof, as we believe that if you are able to go to the doctor's you are able to come to work.

DEATHS (other than your own)- This is no excuse; there is nothing you can do for them so you will have to let someone else take care of the arrangements. However, if the funeral can be held in the late afternoon, we will be glad to let you off one hour early, provided that your work is ahead enough to keep the job going in your absence and you make the time up the previous day.

LEAVE OF ABSENCE- (for an operation) We are no longer allowing this practice. We wish to discourage any thoughts that you may need an operation as we believe as long as you are an employee here, you will need all of whatever you have and you should not consider having anything removed. We hired you as you are and to have anything removed, would certainly make you less than we are paying for.

DEATH (your own)- This will be accepted as an excuse, but we would like a two-week notice as we feel it is your duty to teach someone else your job before you go.

Also, entirely too much time is being spent in the restroom. In the future, we will follow the practice of going in alphabetical order. Those whose names begin with an "A" will go from 8:00 to 8:05, those with a "B" will go from 8:05 to 8:10 and so on.

If you are unable to go at your time, it will be necessary to wait until the next day when your time comes again.

7. A LAYOUT'S LOOK AT WORK

JOHN GEAREN—John Gearen, and a friend of his whom I learned to call "the professor," taught me the differences between the way things are supposed to be and the way they are.

John Gearen[1] is a big man. His six-foot-two-inch frame has filled out, and much of the hair that once crowned his head has disappeared. He has large, work-worn hands. His knowledge of his job and of Western Electric constantly overwhelmed me.

John Gearen is a bawdy man. He is one of the very few men who can get away with telling most women off-color jokes. Some may feign embarrassment, some may even be embarrassed. Yet John can get a laugh from remarks that, if made by other men, would bring looks of horror.

On any given workday John Gearen's mood can make very wide swings. At one moment he can be coaxing production from you, promising to get parts that other layouts would not or could not get. He disappears, off to the storeroom, and returns, only for a moment, with the news that there are no parts in stock. He disappears again only to reappear; that large hand opens and inside are cupped some little capacitors. As I take them he explains that he has gotten them from a friend in another section and he'd take care of him.

On that same day he might come by and tell you, "Don't work so hard kid, they aren't going to pay you for it. Just pace yourself, make up your bogey. Those bastards don't care about us." That was John Gearen's bitter side. He has not gotten the rewards he feels are justly his.

"It burns them up," John says. "They hate it that I don't do any extra here. I'll work my forty hours a week, but no more than that. You know I used to moonlight rather than spend any extra time in here. I told my boss many times, 'I can only stand you five days a week.' And I don't mean because he's vicious. It's because of carrying other people.

"That's what gets you, carrying people, not the work. Well, work never killed anybody, but what they got going down in the shops is

[1] The names in this chapter have been changed.

bad. This is what makes the dissension in the shop. It isn't the physical labor or because you come home tired; it's how many people you have to carry. Oh, does that get to you.

"It's happened to me so many times. There used to be this 211 grade I knew. He'd make out the shop order forms. The job would come to the floor and I'd tell him they were wrong. He'd ask me if I was sure and after a while he just told me to do them my way. He should be the one telling me, he's the higher grade but I'm telling him, see, doing part of his job, and they keep moving him up.

"Hell, there are all kinds of stories about people around here carrying other people. And it's happening more now with the women's lib. They keep promoting women who aren't qualified to satisfy the government. I don't mean there aren't any qualified women, there are. There are some crackerjacks, but they got some of these dames over their heads. The only way they'll keep them going is to put some guy like me around to carry them until they really learn the job.

"Guys like me in here, we're like mustangs, that's something from my Navy days. In the Navy, a mustang was a guy who really knew what was happening but wasn't a commissioned officer. A mustang would bring the new wet-behind-the-ear Annapolis graduates and show them what a boat was about. Not in the classroom, but out in the water. And it's the same in here. The guys who know everything, they can't promote them because they don't think they're smart enough. But they know all the ropes. It's the mustangs here who are helping the guys who have all the titles.

"You know," John continues, "your layout is really your key person. The layout operator can make your supervisor smell like a bunch of roses or a basket of shit. There's so much we can do either to screw them up or to make them look good. I'm not bragging, but I can get the people behind me to work. I can either make them go like crazy or I can walk in back of them and have them all agreeing with me, how they're getting taken and everything, and I'll have them all bent out of shape where none of them are working.

"Most good layouts can do that, they can have quite an effect on a group. Do you think they give you credit for working hard, for getting extra work out of your group? No, they run this place just like Hitler's Germany. They play on your fear, one against the other. They keep telling you keep your nose clean. Then they dangle the carrot, and they keep you on the string as long as possible. Jeez, I can give you all kinds of examples of that. When they're thinking of promoting a layout to supervisor, they call several of you and tell you individually you got a real good chance. Well, you find out they told that same stuff to a half-dozen other guys, and they think you'll kill yourself to look better than the other guy.

"I'm telling you all kinds of stuff like that happens in here. Everybody is watching to get ahead and not step on anybody's toes. Well, once they know that, they've got you. If you've got any gray matter, I'm not talking about being a real brain, just if you've got anything upstairs, it's got to get to you.

"They're just constantly doing things that tick you off. Once, years ago, they came to a department where I was working to take pictures. There was one terrific-looking broad in that department and another plain-looking one who was a terrific worker. Naturally they started taking pictures of the good-looking one. It got me mad, so I really embarrassed them. I asked the photographer, right in front of everybody, 'Why are you taking pictures of her? Why don't you take pictures of this woman, she's a damn good worker?' They couldn't get out of it so he had to take pictures of her and they printed it in the WE Newsletter.

"See, if you get me started I can think of more and more things they do, how they turn people against each other. You know I can go on all day. I'll give you one last example and then let's talk about something else.

"You know the company suggestion program? What they don't tell you is that they don't give you much for what you suggest, not considering what it means in saving for them. Most of the time they just write you a thank-you note. I was suckered on a tool that I designed. I did the whole thing and my supervisor said, 'Don't give them that. You can break the rates if you use that tool. Look, you can get all the work done in the morning and have the rest of the time to sit around.' Like an idiot I listened to him, and I found out later that he went to some national meeting and took my tool design with him. He got a pat on the back for my design and credit for a tool that was in every bay assembler's tool box. Now he's got a good position out in San Ramon [California].

"But that's the way it is around here, everyone looks out for themself. I've helped a lot of guys get ahead. You'd think they'd try to help you out. You go to the top management and you say how about a break—I'm at this dead end right now, how about going out on a limb for me? They always tell you the same thing: we'll keep our eyes open, but at the same time they're pushing people up over you with less experience.

"So you stay in the same place. See, everyone in here has a label. I don't care what it is, you're a troublemaker, you're a company man, you're a union man. I wouldn't put it past them to have a three-by-five card on you up in the personnel office. The label stays with you as long as you stay here. I'm known as starting new jobs, I was only asked two weeks ago to go on a new job, no more money, nothing, just glory, and this is what I don't buy anymore.

"You know, I see it all, and that's the problem. Some of them don't know what's going on in here, but I do. But I can't walk out, I have too much invested. They've got me by the balls. I got time in here. I got security in here. I've got all kinds of benefits. That's how they entice you in here, with social welfare.

"They get these women in here on loans. You know they're in here for the second car, or they owe for a swimming pool; they owe for this or that. Before they know it, they're all tangled up. That's how you get stuck. You talk to people in here, they'll tell you they came for a year or two. But they're still here. They still owe money.

"Maybe when I first came in here I could have left, but I didn't. I guess it was because even though I wasn't old during the Depression, I remember it. I remember my father putting rubber in our soles and I remember being heated by the stove. I wanted more than that. I've taken a lot to get that security. I've given up on getting ahead, because I'm not willing to play their game.

"The only thing I do is limit my time in here. Jack said to me tonight when I was leaving with you, 'What do you say John, coming in in the morning?' I said, 'You got to be kidding.' See, I'd rather do moonlighting and get less because it doesn't take so much out of you.

"I haven't worked a second job for almost five years now. I used to moonlight nearly thirty hours a week working in a liquor store. I'd work a couple of nights a week and then work on Saturdays from eight in the morning to eleven at night with a couple of hours off. But the hours didn't bother me. It's completely different from the factory, working behind the counter.

"I liked the work. You're on the other side of the counter and you're meeting all kinds of people. I can tell you story after story about people. Some of the funniest are about people who work at Western Electric. I've had a couple of department chiefs come in to pick up a couple of bottles and then, when they see me, they never come back because they're not going to let me, working in the shop, know how much booze they drink.

"All kinds come in. I remember this one guy came in looking glum. You can tell he wants to talk so you ask, how's it going? He says, 'I'm having a party down at the house, and there are two other guys and my wife, and they sent me to get some more beer.' That's not bad, I say, trying to act dumb. His face gets a mean look and he whips out some panties from a pocket and says, 'They ain't kidding me. I know what they're doing with my wife while I'm gone.'

"Well, he's one type. Then there's the kind of guy that I get a kick out of. He comes in and says I don't use this stuff myself. I'm buying it for the old man down the hall, but give me a dollar bottle of that wine. Then I get the housewife who comes in in the morning and says give me a bottle of Schenley, a quart, and then she comes back at seven in the evening. Well, she's surprised to see you still working, she gets another quart and tells you the one she got in the morning wasn't for her. You get all types in the liquor store.

"The beauty of a liquor store over a bartender is that they're not there that long. The bartender has them crying at the bar all night long, in the liquor store, they're in and out because the law says you can't open the top of the bottle inside the store.

"There's something going on all the time, and that fascinates me. I never liked the stuff that much. Hell, I think we still got a half-gallon bottle of vodka around that's been in the mountains, been to the beach three times. I never worked the liquor thing to be near the stuff.

"I quit when I got on top of the pile, paid off the mortgage, and put a little away. I stopped about five years ago, I figured I wasn't getting any younger. I didn't need the money, and well, I didn't want to work until ten at night any more."

At this moment the kitchen door opens and Mary Beth, John's wife, walks in.

"When we got married," John says, "we both were working full time. We didn't have a big bank account but we had a couple of thousand bucks. We saw this place and we bought it. You should have seen the place. There was no furnace in here and the squirrels were running through the house. There was no heat, no bathtub. I put a shower right there, right behind you. I added this kitchen on. The house kept me awful busy weekends, doing most of it myself and then hiring a few here and there to do tough work.

"While a lot of our friends were snowed under by mortgages and a couple of kids, we both worked for five years after we got married before Elizabeth came. We had enough money to go away, go out for dinners, fix up the house, and put a little away."

"Did you quit working when your first child was born?"

"No," Mary Beth says, "I took off a couple of months and then I started working evenings. I worked from seven at night to one in the morning. John came seven years after Elizabeth and I didn't go on days again until two years ago."

"Why did you go back to work?"

"At first to just get the vacation, but really we needed it. You couldn't really live on one income."

"Not and have a little extra," says John. "We always had enough to take a nice vacation. I always had a boat, maybe not brand new. We bought an old boat and repaired it, and we've bought land down by the water."

"See," Mary Beth says, "we didn't want to have a baby-sitter. That's why we worked different shifts."

"We didn't go along with what they do today," John says. "Maybe it's old-fashioned, but we figured, we had them, we'll bring them up. I'm the one that put them to bed and gave them the bath. They went in the sink and the tub."

John finds some pictures of how the house used to look. "It's changed a lot, hasn't it?" he asks.

"It sure has," I say.

"I'd still like a new house, though," Mary Beth says.

"We've got everything modern in this house, but Mary Beth says she doesn't have enough closet space," John says.

"I can tell this has been a long-standing discussion," I jokingly remark.

"Oh yeah," John says, "every woman wants a better house. But you can see the house. What I'm paying for taxes and everything, we're living in a very good neighborhood, and living cheaply. Really we've got no business living in this neighborhood. The lots around us are selling eight to ten thousand and up. Everyone else has a big expensive house, not us. I'm the only working stiff in the neighborhood, everybody else is a businessman or executive."

The door opens again and Elizabeth walks in. She and her mother soon leave to do some shopping. John takes me for a tour of the house and I meet Mary Beth's father, who has lived with the Gearens for the past two years.

Island Street is only a five-minute drive from Western Electric. It is not far away from where John Gearen grew up in Haverhill, but a lot has changed. When he was growing up on Terrance Street, Western Electric hadn't been built. The whole area was dominated by the shoe factories. John Gearen's father came as a young man to work in those factories.

"My father came from Ireland and went into the shoe factories. He made box toes, and the counters for the back of the heel. He stayed with it for a while but he got sick of being inside. He went out and became a licensed rigger and a painting contractor. He did that till he got too old, and then he went to work for a public utility company as a janitor.

"My mother took care of the house, and even in the worst of times we were well, because she was a good cook. The Irish aren't famous for their cooking. When my mother came to this country she didn't know how to cook very much. She took a job doing housework for some wealthy family. My mother started off doing housework, making beds and then chambermaid work, and she finally got into the kitchen. There was a German lady that was doing the cooking and she learned to cook from her. My mother is Irish but there'll be a German background to her breads. My mother makes a strudel but it's almost—an Irish strudel.

"It was a different kind of living back then. We lived in a tenement district. You know how it was years ago, a row of tenement houses and everybody on the block related, your aunts and uncles and everybody across the street. We never traveled very much.

"To us the big thing was to get out to Salisbury Beach, which was only sixteen or twenty miles away. Or on weekends the big thing was Hibernian Hall. It was an ethnic club. Every ethnic group had its club and that was a big thing. The women would cook and they'd have their dances. They've all disappeared now. I've got happy memories but there were hard times. Hell, most people had hard times then.

"But I lived better than most. Like I told you, we had food and everything and I never went without clothes. I had a lot of hand-me-downs. See, my mother had a sister that lived in Holyoke [Massachusetts]. She worked for very wealthy people; the kid had all his clothes from Saks Fifth Avenue. And I used to wait for that railway mail truck to show up with a bundle of what the kid didn't want. My aunt used to send it to me. And this was my wardrobe. A lot of it didn't fit but I had spats and patent leather shoes when nobody in town could afford them or even afford to look at them. So I always had clothes and of course my confirmation was the first suit of clothes my parents ever bought me. That was a big occasion. They went for sixteen dollars for a suit and two pair of pants and a vest.

"Another big treat for me, as a child, was to go to my aunt's in Holyoke. Every summer I'd take over this paper route of a boy from up there who went to summer camp. I'd get put on the bus and then on the other end of the line they'd be waiting for me. And then I'd spend summer there. I think that was my biggest treat.

"It may be hard for you to understand growing up with cars all around, but just to go away was a big thing. Like I said, we had no automobile. A streetcar was a big thing, so to make the trip on the Trailways bus, to go what seemed to me was a thousand miles away, was unbelievable.

"I stopped going when I got to be fifteen. That summer I got a job down at the beach. My father's sister, another aunt, got me a job, and I stayed at her rooming house down by the water. She used to give me a cot in the corner with one of these curtains that fold and that's where I slept.

"To me it was great to be working on the beach because I got to meet all kinds of people. The war had started and I can remember French sailors coming in off a submarine. All the big bands used to play down there. I met all those big guys: Jimmy Dorsey, Tommy Dorsey, and Gene Krupa. When Artie Shaw came down with the big forty-three piece orchestra, they had to make the stage bigger. I got his autograph. This was great as a kid.

"This aunt that let me stay with her down at the beach, well she had no children of her own. She used to give me and my sister a special Christmas present. She'd come over Christmas and put two twenty-dollar bills on the Christmas tree, one for me and one for my sister. The next morning, Christmas morning, she'd come over, take the bills off the tree and give them to my mother. They were always used for the coal. Coal was fifteen dollars a ton and the Christmas money took care of it. We never got to touch it, just look.

"You didn't need much money in those days, so I didn't mind. I think the thing I did mind was we never traveled. I wanted to travel, and I quit school when I was sixteen to go in the Navy.

"I signed all the preliminary paperwork before I went in. I went through my induction on my seventeenth birthday in Portsmouth, Rhode Island. I was big for my age, so I was able to get away with stuff because I always seemed older. When I was seventeen I was six foot one. I was sent to school to be a bos'n's mate. I got two stripes right away because of the school. I went in as the top petty officer on the ship. It didn't have a chief. It didn't call for a chief; there were three officers and two petty officers and an acting chief. I was doing the duty of the chief; making liberties, putting guys to bed, waking them up. I had to bluff it a lot because some of these guys were on the Merchant Marine run before we got into the war. There were some tough ones and most of them were older than me. One incident—here I am seventeen, maybe eighteen, years old, I'm telling some guy to rise and shine and he pulls a knife on me. I flip him out of the bunk and then the crew jumped him. They were going to do a job on him, but I told them to leave him alone. He had a tough time, and it was enough.

"War makes you do funny things. I'll tell you I think of Calley sometimes. He was only a young kid and all of a sudden they put a big command under him. And the easiest thing to do is to follow orders; that could have been me. Luckily I didn't have the chance. I spent most of my three and a half years in Europe.

"We went into Omaha Beach that morning. We carried the First Division in. And the First Division was all tough, ready, seasoned soldiers. They were all cool cats; we were just young and nervous and we hit the beach with them and they were great.

"I learned a lot out on that ship. One thing I never liked was the way that if you were a bos'n's mate you were supposed to go out with the other bos'n's mates and not with the seamen. I wouldn't go along with that. When I felt like going ashore, I used to go with this one steward—Geoffrey Lafayette Landen, from Washington, D.C. He was a terrific athlete—a colored boy. The two of us used to take a basketball and go anywhere we could find a hoop when we were in port. I'd rather do that than go out with all those layouts, not layouts, I mean bos'n's mates.

"You know the way the bos'n's mates were reminds me of how things around here are at lunch. You'll see all the layouts sitting together, and the supervisors, with their ties, all sitting at tables by themselves. One won't sit with the other. Go up to the cafeteria, take a look at it, it'll amuse you.

"Anyway, I was glad when I got out of the service. I had seen some places but none of them impressed me. I was ready to come home.

"It wasn't too long after I came home that Western Electric moved up here. Funny, I could have had an early job in Western, been one of the first. See a lot of the guys who came up from Kearney didn't have a bankroll. They weren't getting the pay they're getting today. And they were looking for a place to crawl in. There were two hotels and my aunt ran a boardinghouse. Neil Widett stayed with her. He liked it and told the next guy coming up and soon she was full of Western Electric people. She wanted me to go over to Western. She'd come in our house and say we're going to get John in Western.

"I was working in the shoeshops then, making good money; I was making $125, $150 a week. We worked in pairs, doing piecework. I had a partner and we split the tickets which you got paid by. I'd be all done by two in the afternoon with no dinner hour. You only got paid ten cents on the clock or maybe it was nine cents, so the nine cents you were losing by leaving early didn't make any difference.

"I couldn't see going over to Western where they were making fifty cents an hour."

The door once again opens and Mary Beth and Elizabeth, laden with shopping bags, return.

"Are you two still at it?" Mary Beth asks.

"We're almost done," I say. "I was just going to ask John how he finally decided to move over to Western."

"I think," John says, "that Mary Beth was instrumental in getting me to go to work for Western, weren't you?"

"You were laid off in the shoeshop, John," she says.

"Yeah," he says, "all of a sudden things got bad in the shoeshops. Mary Beth and I talked it over and she convinced me to go over to Western and make an application. That was eighteen years ago."

"Are you both looking forward to retiring?" I ask.

"Well," Mary Beth says, "I'm going to go to fifty-five if I live to that."

John says, "We'd both like to retire when we're fifty-five. What we want to do is to get a piece of land in Florida for the three months that we're not up here. We'll sell this property, but for nine months we'll live in the beach house I'm converting. It will have electric heat, a cellar, two-car garage underneath it and two-hundred-amp service coming in. This summer we're going for a couple of thousand dollars for wall-to-wall carpeting."

The door opens again and John Jr. walks in, tossing a ball. He asks his father if he wants to have a catch. John looks at me to make certain we're done. I smile, shrug my shoulders, and nod my head.

"Well, then," John says, "let's go have a catch."

And the three of us walk out into the back yard and begin throwing the ball around.

8. QUALITY CONTROL

Western Electric takes a great deal of pride in the high quality of its products. At Merrimack Valley there is an elaborate and complicated system of quality control, beginning with the process checkers who check the work done by the bench hands. Defects not caught by the process checkers are picked up by shop inspection, the second stage quality control people. The company would like all defects to be picked up either at the first or second stage. There is, however, one more check before the product leaves the plant, made by company employees who act as representatives of the customer. They are part of a company headquarters organization called "Quality Assurance" and operate in areas called the "cage." A defect that is picked up in the cage is considered quite serious.

During my orientation the need to produce something well was repeatedly mentioned. Again and again we were told it was our responsibility to make a good product. One of the men running the orientation said, "We have a top notch quality control system, but you can't inspect quality into a product . . . you must build it in."

Our supervisor also stressed quality and made a point of telling new workers that it was better to build something right and build it slowly than to build it fast and build it wrong. Although I was aware of the importance of quality, my first week was preoccupied with learning my job.

I was, I knew, working slower than other people, but I was sure each time I built a board that I had built it correctly. Yet, to my surprise, I was being told by the process checker, who occasionally came by, that I was building things incorrectly—reversing polarities, inserting wrong parts, and on two occasions forgetting to insert required parts. The process checker would return my boards. I'd correct the mistakes and resubmit the boards. I thought that that was all there was to the system.

Near the end of the second week I noticed that there was a sheet, displaying a graph, in front of my working position. I looked around and noticed that there were similar graphs in front of everyone's positions. Taking a closer look at the graph, I noticed a normal curve and then another curve, which I assumed represented my productivity. Out of curiosity I plucked my graph from its stand, turned to the

PRODUCT QUALITY
Position Charts

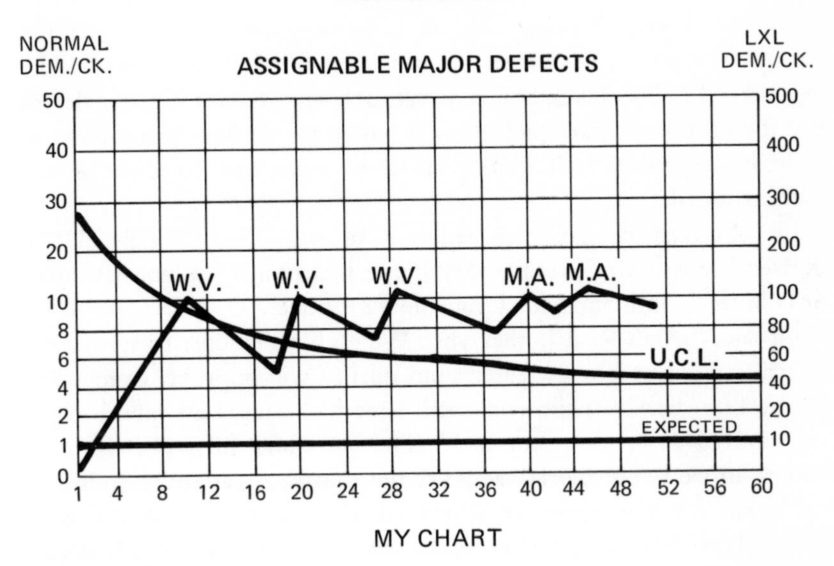

MY CHART

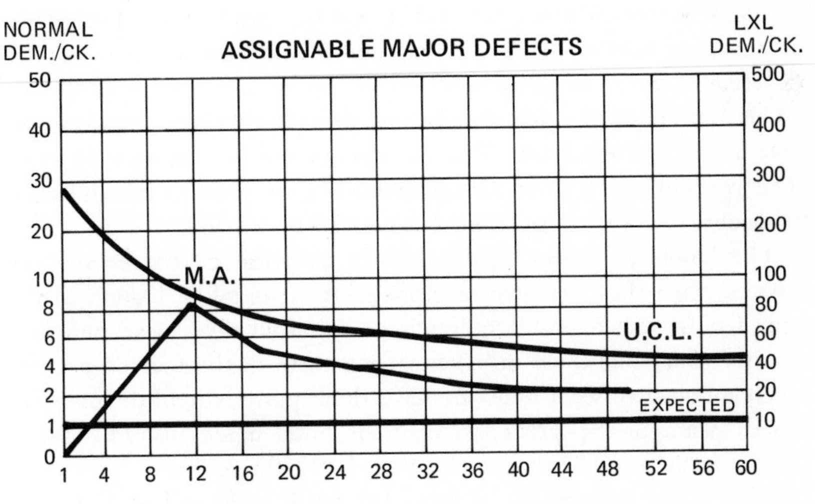

ANOTHER WORKER'S CHART

M.A. = Missing Apparatus
W.V. = Wrong Value
U.C.L. = Under Control Limits

woman who worked behind me, and asked what these graphs were all about.

Jill[1] told me it was a quality control chart. She said the big steady line represented the number of mistakes you are "allowed" to make in a month. "A good worker," she said, "is supposed to stay under the curve." The other line, which on my chart looked like a roller coaster, represented my mistakes. She said that each time I make a major defect that the process checker picked up, it would be marked on my chart as a hundred demerits (fifties are marked as well), and my chart would reflect those defects.

Jill observed my alarmed reaction to my chart, which clearly showed I was having trouble. My curve had passed the normal curve early in the month and hadn't gone under it since.

Jill said, "Don't let it bother you. We all have trouble to start with. When you get to know the job, your quality will improve. Even when you've been here a while you'll still have bad days. Last month I got six hundred demerits in one day." The last thing Jill said before we both returned to our work was, "Try not to pay any attention to it."

I began working again but found myself taking an occasional peek at the chart. Now that I knew what the chart represented I was embarrassed. Somehow it seemed that it was like displaying a low grade for everyone else to see. I couldn't understand how I could *not* be bothered by it.

The next morning when Arlene, the process checker, came over to make a check, I asked her about the system and her feelings about it. "Theoretically, it's a good idea. Obviously they need to try and control quality, but I think the thing has gotten out of hand.

"The trouble," Arlene continued, "is with the charts. People get nervous when they see they've made a lot of mistakes. I don't blame them. I don't think most people mind being told they make mistakes, but they don't like those mistakes staring them in the face on a chart which anyone can see. I know myself I don't put down all the mistakes I catch. Some of the girls chart them all, but I don't. Instead, I bring the boards back and let the bench people know the problem.

"Look at your chart—you're upset because it looks so bad. If I plotted all the mistakes you made, you would be off the chart. But how is that going to help you do better work? It isn't. My trying to explain a mistake or how you should work may help.

"I'll tell you something else. I can tell when one of the girls is having

[1] The names in this chapter have been changed.

personal problems. I don't mean a little argument, but if there's something wrong with her children, or if she had a fight with her husband, it always shows up in the quality of the work. Their minds just aren't on the work. I try to be sensitive to it and I end up being a mother confessor.

"We're having a problem with quality because we do so many different jobs. If you have the kind of shop where you're doing a repetitive task, then you can keep good quality and high production. I worked in an area where we did the same thing week in and week out. We put out ten thousand parts in a week. The girls got so they could do them blindfolded. When you're doing a lot of different pieces, switching around like we do, there tend to be more mistakes unless you give up on high production.

"Really, if you ask me, if the company were smart they would get rid of those darn charts. They just make people feel bad. Most people in here want to do their job correctly, and the company should work under that premise."

As the months passed I heard more and more people complain about the charts. The complaints seemed to intensify with news of our poor bonus. People would say, "Sure we can't make a good bonus, we're always building boards short; you can't get a routine going. We have to do work over, fix mistakes, and they don't pay you for the second time you do a job."

The complaints about quality control and the process checkers mushroomed into real hostility when the group learned that it had gone "out of control." A group is considered out of control when the number of demerits exceeds a predetermined statistical limit based on production levels. This is a very serious situation, so serious that Western Electric's New York headquarters is notified of the problem. The situation is rare and is an embarrassment for the supervisors responsible for the area. A group that is out of control is, I soon learned, slated for meetings and other procedures to locate the group's problems.

The first tangible result of our going out of control was the disappearance of the courtesy check. Customarily, at least in our shop, when a worker began on a new board, he was given one courtesy check by the process checker. If a defect was picked up it wasn't charted. Instead he was simply told of his mistake. In this way he could become acquainted with a new job and not worry so much about defects.

I discovered that we were losing the courtesy check when one of the women started complaining about it during lunch. "Do you know," she said, "Arlene gave Tina three hundred demerits on a new board. Arlene told her, 'No more courtesy checks because our group is out of control.'"

Janet, Tina's friend, was really upset. "They've always given us a courtesy check on a new job," she said. "This new stupid idea will just make everyone more nervous."

Later that afternoon I learned that more than just the courtesy check had disappeared. Arlene brought back one of my boards with three defects, which meant three hundred demerits. Usually in a situation like that she would chart me for only one defect. I felt somehow abused.

I asked her in a half-joking, half-serious way why she had charted me for all three mistakes. For the first time since I'd known her she spoke to me with an edge in her voice. "Look," she said, "my boss keeps coming down on us [the process checkers]. He tells us we aren't charting everything, and it makes it look like we're not doing our job. So from now on he wants us to put down every mistake we catch. They want to know what the work coming out of here is really like."

Tempers grew short very quickly. Boards that had once been returned by a process checker with instructions to fix it but without a demerit being charted were now returned with each and every defect charted. People started reacting negatively to the process checkers, making snide remarks to them and behind their backs. The process checkers retaliated by starting to give the boards with mistakes to Jack, our supervisor, rather than directly to us. It cut down on personal contact and it made Jack even more aware of our mistakes. It was a very uncomfortable situation for everyone.

Jack finally said something when two operators, with more than seventeen years of service between them, had received a total of seven hundred demerits on new boards. With one nearly in tears, both complained bitterly to Jack about the absence of the courtesy check. Jack called the process checker over and asked her to give them a courtesy check. Her face turned red, and she said, "You just told us there were no more courtesy checks."

Jack, as always, kept his composure and said, "Well, I've changed my mind, let's have one."

"No," the process checker said, "I can't."

So Jack went to her supervisor and told him that he wanted courtesy checks reinstituted. They discussed the matter, then Harold agreed and assured Jack that he would take care of it. Later that day the courtesy check was reinstituted.

This made the people on the bench somewhat happier but infuriated the three process checkers. One of them, Vivian, told me, "Here they take one stand and then you do what they tell you and they want to change it. Now that we've begun to chart everything they're embarrassed because there are so many mistakes being made. Either you say there can be courtesy checks or you don't.

"You know what they're doing, they're turning the girls on the bench against us. They don't get mad at the supervisors, they get mad at us, and for what?—doing our job. You think I like charting all the mistakes? I don't; none of us do. We try to catch the mistakes and get them corrected. I don't care whether none of the mistakes are charted. I think it's a dumb system anyway. But we don't make the rules, the bosses do. If I have to chart everything, I will. And I'll tell you this, if the girls start giving me a tough time because I'm doing my job, I'll just give all the boards with mistakes to Jack. We'll see how they like that."

Relations between the process checkers and the people on the bench continued to deteriorate. Everyone was concerned that the bonus was low, so people were pushing to get more work out. But the more we pushed, the more defects we made. Tempers flared. One fellow who had worked for the company for less than six months refused to let a process checker take a sample board he was working on. He told her he wasn't ready. They got in a huge argument, which ended in both of them trying to grab the board.

At the end of what had been a particularly tense week Jack told us we were going to have a meeting with the quality control people to try to figure out the troubles we were having with production and quality. Jill waited for Jack to leave and said, "Big deal, we're going to have a meeting. What do you think a meeting will help to do? All that will happen, if we say anything, is we'll get in trouble."

Another person said, "You're wrong, they really want to know."

Jill interrupted, "Want to know what? Do you think they're going to admit they're doing wrong upstairs? Not a chance. I'm telling you if we say something we'll only hang ourselves. I know what happens at these meetings. They won't do anything about our complaints. They just want us to see things their way. There's no way I'm going to say anything."

The meeting was held the next morning in an upstairs conference room. At 8:30 everybody put down their tools and walked slowly upstairs. The engineers and time study men were already in the room. Our supervisor and our department chief also attended the meeting. The quality control engineer for our department, a person few of us had ever met, chaired the meeting.

Ralph began by saying, "I'd like to be bold enough to make some suggestions on how we can help you people improve the quality of your work. Quality is important because it's important to our customers." Seizing a pamphlet, he continued, "See this, it's a quality bulletin. It is used by all our major customers. If the quality isn't good the customer doesn't want the product. Western Electric has always provided the Bell System with what it needed, but the telephone companies aren't obligated to buy from us. I'll be honest with you: five years ago we weren't in such a competitive position. But today there is unbelievable competition from other companies in this country and from the Japanese. Now if we can't build something well and build it inexpensively, the Bell System Companies will buy elsewhere.

"Building something well means good quality. We have constructed an elaborate structure to ensure high quality. Good work from you people on the benches is important. After the work leaves your hands we have a system for checking the work. The process checker is the first level of that system. We keep making checks all along and the final check is made by the people in the cage. They are supposed to be the customer's advisers. So we don't like mistakes to get that far. If a mistake is made in here and it isn't caught until the product is installed in the field, that mistake can cost the company millions and millions of dollars.

"I don't know how many of you ever heard of Pareto, the philosopher, but he once said 20 per cent of the people own 80 per cent of the land. Well, in here we have our own Pareto principle of defects. We say 20 per cent of the people make about 80 per cent of the mistakes. We use the quality control charts to isolate the cause of the mistakes. By using the charts we can pinpoint where mistakes are coming from and correct them. I know that some of you don't like these charts, but you shouldn't take them personally. We only use them to help us in isolating mistakes.

"Well, I've talked for a long time. I guess that's pretty much of what I want to say. Now I wonder if any of you have any questions?"

After a moment of silence Jane, a person who had worked in our shop for less than two months, spoke up. She said, "Well, yes. I find the quality charts to be very discouraging and humiliating."

Even before Ralph got the chance to get the why out of his mouth, Jane continued, her voice full of emotion. "If I see the chart one more time, I think I'm going to scream. I've worked for several companies and I've worked a long time for this company and this is the first time I've ever been in this situation. I keep on making mistakes on this new job. Every day I make a mistake and I see my chart and I'm afraid I'm going to do worse and I get so nervous I do worse. I hate that chart staring me in the face. I don't mind my supervisor seeing it, but I don't think it should be up for everyone to see."

Ralph tried to mollify her. He said he could understand her being upset but believed the charts needed to be up. Ralph said, "Those charts are up there not to embarrass you, but so the supervisor can see them. We're afraid if they were just given to the supervisor they might get buried in his desk."

I waited for someone to say something. I was sure someone would point out that our supervisor rarely if ever went around checking the sheets, so that couldn't be the reason to have the charts up. Since no one said anything, I finally made the point.

Ralph smiled and repeated what he had just said. When Jane started complaining again about the humiliation she felt, he looked at her and in a lowered voice said, "Maybe there's nothing to say except that you aren't working well."

The room was silent again and then Grace, a woman with eleven years of service, said, "You know, as long as I've worked here I've never understood the need for those personalized charts. I remember when I began working here I was always embarrassed by them. If you ask me they aren't helpful; they just make you feel uncomfortable. I don't know why the company uses them."

Larry McNeil, our department chief, spoke for the first time. "Why do we need them? Of course we need them. How are you going to tell the bad apple in the bunch unless you have the quality charts?"

Ralph said, "I feel you people are taking this all too personally. Those charts are posted to help you, to help you improve your work. Look, if you know what you're doing wrong, you can improve on it. If you make work with fewer mistakes your bonus will go up."

I said, "If the quality control sheets are so helpful, why aren't the ratings of the supervisors posted to help them?" People laughed.

Ralph, instead of answering my question, asked if there were any other questions. When none were put forth, he closed the meeting.

We walked back to work, and for the rest of the day people were talking about the meeting. Jill, the woman who had warned me to be quiet, said, "See, what did I tell you? They didn't want to hear what we had to say. They had an answer for everything, didn't they? They're not going to change anything. That's why I didn't say anything. I felt sorry for Jane, but she should have known better. She's been here long enough to know not to speak up at one of these meetings. You'll see, they're going to make it even harder on her."

To my surprise, a number of people were critical of Jane. Like Jill, a number of people thought Jane should have known better than to speak up. One woman said, "If my work was as bad as Jane's, I'd be upset. But the charts don't bother me; in fact, I like them. I don't make hardly any mistakes and I like to see that my chart is better than most of the other people. I'll tell you," she said, "I like it just like I like the bogey. I do good work, and I like some credit for that."

A few people thought Jane had been courageous to speak up. One fellow told me, "I didn't say nothing because I don't speak English so good, but I feel just like Jane. But now I'm glad I didn't say nothing because all they do is bullshit with you when you tell them how you feel."

I asked Grace why she had spoken up. She said, "I know it won't do me any good, but it just got me mad the way they were picking on that poor girl. I just got madder and madder till I had to say something. I've been here long enough so they can't do anything to me, but Jane could get hurt speaking up like that."

Things didn't change very much in the next month. There were several more meetings, but not with the entire group. There were meetings with the layout operators and the process checkers. Our supervisor went to meetings, and quality control people came down to the shop. We were checked more frequently by the process checkers, and an additional process checker was brought into the shop.

Our bonus didn't pick up. Relations between the process checkers and the people on the bench continued to be strained. Our courtesy check had disappeared again. The process checkers used Jack as an intermediary to pass back our defects until the people on the bench stopped complaining.

I asked Arlene how long she thought the tensions would last. She said, "I'm afraid it will just get worse and worse until the bonus begins going up. Some of this is just mass hysteria. Once people start making mistakes, they begin getting a little hysterical. Not consciously, but it happens. I know, I've seen it before. I can't explain it to you, but mistakes seem to be compounded.

"When a group is out of control like this one is, things seem to get worse for a while. Feelings are strained and then, I don't know why, they'll start slowly to get better again."

There was no big event that I could see that began to change things, but by the time I left some of the tension was undeniably gone. The fairly good relationships that had existed between the bench people and the process checkers seemed to have been renewed. It appeared as though the old practice of not charting all defects had also returned.

At a time when there is increasing national concern about the deteriorating quality of much of our manufactured products, Western Electric takes great pride in the continuing high quality of what it produces. Much of the credit must go to the complicated and elaborate quality control system.

The quality control charts are an important part of the system in that they are used to pinpoint defects. Most people were willing to admit their mistakes, but few people wanted their defects constantly staring them in the face, open for everyone else to see. Many people mentioned that the charts made them nervous. The public display of the charts is degrading and, I believe, an ineffective way of trying to improve quality.

The existence of those public charts, and the reaction of the people on the bench to them, creates a dilemma for the process checkers. They receive conflicting pressures from management and the people on the bench. Management instructs them to chart each and every defect. However, in their daily contacts with the people on the bench, the checkers see the high human costs involved in charting all defects.

It is interesting to note how the process checkers in our department resolved this dilemma. Basically, the three of them evolved a system that varied with their respective personalities but that boiled down to not charting all defects.

Their aim was to catch defects and to see that they were corrected, but they were not intent on charting all defects. None of them seemed to like the charts particularly. In addition, they seemed very sensitive to how the public display of the charts affected the people. Repeatedly they mentioned that they saw their job as *catching mistakes* and not *charting people*. As one of them said, "If you've ever worked on the bench and had to stare at one of those damn charts you know how nervous they can make you."

It was only when our department went "out of control" that the process checkers were forced back into the position of charting each and every defect. None of them believed this would help cut down on defects. However, if that was what the bosses wanted, that was what they'd do. When they began charting everything, they felt the increased displeasure of the bench people who had gotten used to the more informal system.

Their very human reaction to this hostility was to retreat to an even more formalized procedure. Their giving boards with defects to the supervisor and not directly to the bench people reduced personal contact. The bench hands knew that if they didn't co-operate the process checker could make things worse. There is no doubt that it was within the power of the process checker to make a bench worker look better or worse.

But the process checkers were as unhappy with the more formal structure as the bench workers. As soon as complaints lessened, charting began returning to the old pattern.

The stated purpose of the charts is to pinpoint defects and help improve quality. One would think that, when the deleterious aspects of such a small feature as hanging the charts had been brought to the management's attention, it would have been corrected, but that isn't what happened. Lower management ignored the feelings of the bench workers and, what's worse, gave credence to the feeling in our shop that speaking up at meetings was a waste of time.

Instead of trying to understand why people were so upset by the charts, the management's representatives told people they were taking the charts too personally. Of course a person would take that sort of chart personally. Imagine if your public school report cards were posted in front of your desk for the entire year so you could improve your work. Certainly a few people with excellent report cards wouldn't mind. However, we can imagine what the posting would do to others. Or imagine if the practice on the shop floor were extended to cover the ratings of supervisory personnel or engineers. All these people are rated, each of them may know his rating, but how would they respond to a public display of those ratings?

Workers are told during orientation that you can't process check quality into a product. This is true. For the company to function, the quality really has to be a constant concern of the people on the bench as well as the engineers and professionals. The company must act on the belief that the people who work on the bench care about what they build. They should be more sensitive to the human problems that their quality control program has created.

9. THE BONUS SYSTEM

We are a society committed to high productivity. We still haven't found an acceptable way to measure a teacher's or a lawyer's productivity, although lately there have been some unsuccessful attempts to bring the time study approach to municipal workers. However, it is primarily the world of the blue collar worker in which output is directly measured. On the assembly line the speed of the line controls productivity. For others, whose reaches, thrusts, and bends have been measured, piece rates are set and salaries based on productivity. At Western Electric there is a complicated system that combines an hourly wage with a bonus system based on rates and group productivity.

Before I even started work, I was told that I would enter a group and, depending on that group's performance, would receive a bonus. The bonus and how it was determined were things I was anxious to find out about.

During the first days of training, rates and the bonus were frequently discussed. One of the women in my training session had twelve years of service with the company. She was being transferred to another job and was relearning some basic skills. She took it upon herself to tell us about the bonus system.

"In here," she said, "the company will tell you that the productivity of your group will determine your bonus. Well," she said, "let me tell you the company will try to keep the bonus between 24 per cent and 27 per cent. You'll see most groups in here make somewhere in that range. If your group starts making much more than that or if you're making considerably less, there's probably something wrong with the rates. You'll see," she told one of the girls, "they'll keep on adjusting the rates until the bonus falls in that middle 20's range [the plant average is 25 per cent].

"Sometimes," she continued, "you get a bonus which is way out of whack. It's usually because of the rate that a time study guy has made. Like I knew this one guy, we call guys like him Santas. He's an industrial engineer, and he sets very easy rates, rates where the group doesn't have to push itself to make a high bonus. I knew this guy pretty well. He was people oriented. His mother used to work in a plant, and he felt people had it tough, and he wasn't going to set very stiff rates. Anyway, after a couple of months he was switched to another job, and the company brought in a very tough person, which is something they typically do when they think the rates are too loose. He recalculated almost every rate in the shop and the bonus fell immediately back where the company likes it."

That was the only view of the bonus system I had until I joined my work group. I was immediately curious to know what our group bonus was and what the rates were on the boards I am going to make. I was too uncomfortable with the new work to ask. The first day I was having a tough time just trying to do the job my supervisor had shown me. I didn't have to wait long, however, to find out about the bonus. The second day of work our supervisor called the group together to discuss the last month's bonus.

Up to the last month my group had been paid a bonus of 20 per cent. Ours was a new job. At Merrimack Valley on certain new jobs the workers, for periods up to eighteen months, are not given a bonus based solely on what they produce. Our particular group received a 20 per cent bonus for a year. The year was up and for the next several months we would be receiving a bonus that would combine what we earned with a managerial allowance. Finally we would move to the system where the bonus would be based solely on productivity.

Jack,[1] our supervisor, called us all together and announced that the group had earned a 6.2 per cent bonus, but with the addition of a managerial would again get 20 per cent. Since this was what everyone had received for the past twelve months, I was surprised at how angry Jack's remarks seemed to make people. Several women said that they couldn't believe that's all they had earned.

Jack tried to calm them by saying that he wouldn't be discouraged by that figure. He said, "I've never been in a group where the bonus had come in positive the first month off the managerial. In fact, I've been in groups where it's been as low as minus 20 per cent. I'm sure," he continued, "that in the coming months the 6.2 per cent will edge up to a much higher figure." Jack ended the meeting with that statement.

For the next few weeks several people complained about the bonus. One woman, who had surplused into this group, told me, "The last group I worked in I made a 30 per cent bonus. I didn't like the job, or the boss, but we made a darn good bonus. I like it over here but not for this kind of bonus."

[1] The names in this chapter have been changed.

I kept hearing this type of remark. People began complaining that the rates must be too tight. I was still curious about the rates. When I felt somewhat more confident about my work, I asked Jack the rates on a board I had been working on for the last few days. Jack looked through his rate book and told me the rate was six boards an hour. I didn't say anything because I was making two boards an hour. That afternoon I took only one bathroom break, rather than my usual two, to try to increase my production, but I was still averaging about two boards an hour.

The next day I told Linda how much trouble I was having even coming close to the rate. She said, "Don't ever come close to it, not until you've been here a long time. If you're doing the rate after being here for six months," she said, "then they'll think you can really do twice the rate once you've been here for a couple of years. I never made the rate when I started working here. I didn't care. Now I've been here a couple of years, and I can make the rates on most of the boards. But I know if I began going over the rate too much they'd just pick the rate up on me."

I wondered how some of the other women felt about the rates, so later, during lunch break, I told both Rita and Joan, both of whom had more than fifteen years of service, how I had worked so hard and hadn't come close to making the rate. They both told me not to worry. Rita said, "You know, in all the years I've been here I've never asked a rate. I do what I can, and if they don't like it, they can show me how to do better. I know how quick I am, and I'm as quick as anybody around."

Joan said, "I put in a decent day's work. If they don't like it, they can tell me, 'There's the door,' and I'll be out it. Anyway, I'm not going to put everything into the company–I'm saving a little for my old age."

I found in the next week or two that the two other men who started work when I did were also finding out about rates and discovering that they weren't coming close to them. We talked to some of the older men about it. When Arnold said he wasn't coming close to making one of the rates, Jerry said, "What do you care? Don't ask for the rates. Just do your job, nice and steady. Don't kill yourself, they aren't paying you no six dollars an hour."

The next day Helen Guzzi, a process checker with nearly twenty years of service, told me, "When I first worked here I really thought the rates meant something. People would come and study your work for a few days and then they'd set rates. It seemed in those days that if you worked fairly diligently, you could make the rate. And if you really concentrated you could break the rates.

"But it's different now. They don't even need to send people down to watch you. They calculate certain motions that each board requires and they compute them up in some office and come out with a rate. It's not a very human system.

"I don't think," she continued, "people believe the rates they set have anything to do with what we're doing. People don't think the rates, or for that matter their performance, is connected with what they earn for bonus money. Now I can't prove anything," she said, "but I have a friend who knows an engineer and he told her that they compute bonuses up to two years in advance, so it really doesn't matter what you do."

When July's bonus was announced during the first week of August, I had been working at Western for nearly two months. In that time I hadn't met a person who worked on the floor who believed that the bonus system was directly tied to production. I was somewhat surprised, therefore, at the anger the announcement of July's bonus provoked.

Jack called us together to tell us we had earned 5 per cent this past month and that we would be given a 13 per cent managerial, and thus be paid an 18 per cent bonus. I could understand that people were upset because they hadn't made more of a bonus and were therefore losing money. What was surprising was how personally a number of people took this announcement.

One woman said, "I'm not going to do another thing today. I've earned my 5 per cent. This is ridiculous." She said, repeating herself, "Hell, I'm not going to work anymore."

When other people made similar remarks, Jack tried to explain that in the kind of miscellaneous work we did it was hard to make a high bonus at first. He said he was sure that with some sincere application—a phrase we were often to hear repeated—the bonus would begin to climb.

People went back to their workbenches, but work slowed down noticeably for the rest of the day, as people kept on grumbling. Several women with long service complained that they were working hard and earning only 18 per cent bonuses, while some people in other groups with less than six months were making more than 27 per cent.

In the next few days people had a variety of reactions to the bonus, none of them very positive. Fred Hatch said, "You know, they come down here and they say if the guys don't fool around with the girls so much and if the girls don't spend so much time in the bathroom, the bonuses would go up. That's a lot of crap. You can spend equal time working in two different departments and the bonus would be different by several percentages."

Ann Traeger said, "It isn't fair for everyone to receive the same bonus in the same group. Either everybody in the plant should get the same bonus or there should be a piece rate system. I've worked in the mills and in a shoe factory, and people got paid for what they did. Those that worked less got less, those that worked more got more. I do good work in here and I'm penalized. It isn't fair."

I felt everyone's frustration, but I didn't really have a reaction until the next week, when we received our bonus checks. July was a short working month, since the plant was closed for two weeks for vacation, so everyone's bonuses were low. Still, I was disappointed when my check arrived and I received only $4.42. Another worker quickly explained to me that new workers didn't fully share in the bonus. Instead, we were on some complicated learning curve, and for the first ten weeks were paid some porportion of the earned bonus and didn't share in any managerial allowance. I had worked fairly hard the first two months, and the first tangible result that I could see was $4.42.

I guess what burned me and the two other new workers was that other new workers who had started when we did had gotten into groups earning 26, 27, and 28 per cent bonuses. They were sharing in those bonuses while we were sharing in a 5 per cent bonus. I now understood how people felt. I felt insulted that all we had earned was 5 per cent.

People kept complaining about the low bonus, so Jack called the group together. He told us he was disappointed with the bonus because he knew we were working hard. He promised that certain of the rates would be checked and re-evaluated. Then he said, "I think one of the big problems is that you people aren't filling out your bogies correctly." (Bogies are a personal record of the week's work. Each day you put down how many boards you've built, plus time spent waiting, setting up, and material handling. The week's bogies allow the supervisor to determine an efficiency rating that supposedly gives each person a sense of how well he or she is doing.)

Jack said, "I think many of you aren't putting down all your waiting time, or setup time." He looked at JoAnne and said, "Last week I know you went to a conference for an hour, and I didn't see that on your bogey. We have to begin putting down all the time we can get credit for when you aren't producing." Several women agreed that they hadn't put down everything they could charge on the bogies.

After the meeting one woman told me, "See, if we can't make our bonus with production, then we'll have to make it a pencil bonus." I asked Fred Hatch what Joan meant. He told me, "Look, we get certain credits for nonproduction activities like waiting time. Jack wants us to mark it all down. You'll learn to do more than Jack asks, to pad your bogey, everyone does. If it takes you thirty minutes to set up, mark it down as forty-five. If you finish a job in forty minutes, put down forty minutes for the job and another twenty for material handling."

That was the first I had heard of padding a bogey. Until then I hadn't taken the bogey too seriously, nor had my friend Mario. Neither of us had moved our pencils and our efficiency rates were below 50 per cent. Mario told me that he had hand inserted more than 1500 parts one day the previous week. He couldn't believe that anyone could do it faster. We both agreed that, starting then, we were going to do as the older workers had suggested: mark everything down on our bogies.

The following week I began to put down all the time I took doing set up, or waiting, or material handling. My efficiency went up to 73 per cent and the following week to 84 per cent. Even though I hadn't worked harder I was surprised to see how much better I felt because my efficiency rate was up. I discovered that most people felt better as their efficiency ratings went up, even if they weren't working differently, but using a pencil to its best advantage.

Within two weeks of our group meeting, the time study people began to come around for their first rounds of re-evaluating the rates. One day they watched Claire as she worked on a board. After they left she told me. "You see, I work steady, but I'm not going to kill myself, because once they set the rate you have to live with it."

That's what everyone said: don't work too hard when the time study people come, don't kill yourself, don't show them any short cuts, just work steady. Fred put it in stronger terms. He said, "I told those broads not to talk to those time study men. All they want to do is use us. I tell them if you want to talk to them, you might as well just open up your pocketbooks and let them take your money, because they want to figure out ways to cut down further and further on your rate and why the hell should we help them do that?"

One of the older men told me, "If those bastards come by don't push yourself. Just do your job, and remember, just do one board at a time. [Rates are set on doing one board at a time, even though people frequently work on three or four boards at a time.] If you start impressing those guys with all the short cuts you've learned, they'll just push the rate up on you."

I was determined, if a time study man came to watch me, to co-operate as little as possible. What did I have to lose? When a time study man finally came by and began watching me, I found myself getting very nervous. His presence made me uncomfortable. Instead of taking my time, I worked hard. I know I was trying to impress him.

There was a high level of anxiety in the shop until the end of the month. After the afternoon break on August 31 everyone was crowded around Jack's desk. I walked over and heard that we had moved our earned bonus up to 10.3 per cent this past month, but we were only going to be given a 7.2 per cent managerial. Thus the month's bonus was 17.5 per cent, down from last month's 18 per cent.

I could hardly believe what I was hearing. Didn't the people upstairs who made the bonuses have any sense at all? We had supposedly doubled our earned bonus from last month, and as a reward they were cutting the managerial so we would make less. I thought this development placed Jack in an impossible situation. As the company's representative, he had to try to make sense out of something that escaped rational explanation. Surely the company could not intend to reward increased productivity with a declining bonus.

Jack could see the anger and hostility that were building up, and wisely took us up to the cafeteria for a forty-five-minute meeting. "Look," Jack began, "if we can increase by 5.3 per cent this month, maybe we can increase by 7 per cent next month and we'll just keep heading up till we're making a good bonus." That didn't work. People just said, "Heading up where? No matter what we produce, we just keep losing money."

Jack quickly shifted gears. He agreed that it was true that some people were producing well, but not everyone was. Without attaching anyone's name to the figures he read some efficiency rates computed from last week's bogies, ranging from 53 to 102 per cent. Then he tried to put some group pressure on those producing at a low level by saying, "In a small group like this, if just three people don't do as well as they can, then your bonus is going to be hurt, much more than in a big group."

Before anyone could say anything, he said, "Haven't some of you worked harder these last few weeks? I know you must have, because we keep on getting better and better. Your bogies are better and better."

I was sure that someone would say something here about the bonus going down when production was up, but no one did. Instead, to my surprise, a couple of women agreed that their bogies had improved.

Jack seized upon this agreement and ended the meeting by stressing how important it was that we all kept marking things down. He said he was proud of some people and wanted to help other people make it up to what they could do. He then indicated that he would like to speak to each of us individually in the next couple of days.

Whatever personal anger had been dissipated at the meeting and in Jack's personal sessions, it crept back into the life of the shop in the next few days. People were really upset. Several women talked about obtaining laterals to get out of this shop and into other shops with higher bonuses.

In the next couple of weeks arguments within the shop increased. People began complaining about the process checkers and the layouts. There were even complaints about Jack, who up to then had generally been regarded as one of the plant's best supervisors. Even good-natured people like Penny Taylor seemed on edge. One day as another woman passed by she said, a hostile tone in her voice, "I don't get up like some of these people. I sit here and I do my work. If people would stay in their own seats maybe we could make a decent bonus."

Repeated poor bonuses had stirred the group to a point where people were turning on each other. I felt particularly hostile the next week when I discovered that another new worker, Eric, was receiving a bigger bonus check than I for the last month because his bogies were repeatedly higher. I knew that his percentages were higher not because he was more efficient but because he pushed his pencil, padding his bogey more than I did.

Until then I had tried to limit the time I put down to actual time spent away from work. I decided, why bother. I might as well pad my bogey because there was certainly no reward for keeping an accurate record of what was actually being done. In the next few weeks my efficiency rating kept going up. My efficiency went first into the high 80's and then in the 90 per cent range.

The anger that the group increasingly turned upon itself remained until I left. Although our bogies kept improving, our bonus didn't markedly increase. The last month, just before I left for another job in the plant, Jack didn't even take us through a big meeting to discuss the bonus. He just told everyone what the bonus was and left it at that.

The bonus-rate system is very important both to the company and to the workers. Obviously the company wants to get as much production as possible out of workers. Instead of paying a straight salary, it attempts to stimulate extra production by paying a salary and a bonus for increased production. The company has a complicated system on which the bonus is calculated, a system based fundamentally on piece rate output compared to hours worked.

Ideally the company's system may have real integrity. It may allow the company to get increased productivity and it may reward people for that increased productivity. However, because of the way the system is operated and the almost total lack of understanding of the system on the shop floor, the system has minimal operational significance. It doesn't matter what fancy or complicated explanation the company can give for the system—people don't believe it. In the five months I worked in the shop, none of the people with whom I talked accepted that his bonus was based purely on production.

I was told by several different people that the company would play with the bonus, especially in a recession period. "You'll see," someone said, "if times get tough our bonuses will start sliding downward." Another worker said, "That's right, and the only group they don't mess with are the people in the storerooms. They get a 30 per cent bonus no matter what. The company knows that if it gets the people up there too mad they can mess everything up. All they have to do is start mixing the parts bins. But the rest of us, they fool around with our bonuses as much as they want."

This is unfortunate but is part of the general distrust reflected by a *we-they* feeling. The bonus system is looked upon as a system used by management to manipulate higher production. *They* always want more production and are unwilling to pay for it. *We* have to protect ourselves, and that means padding bogies and not helping the time study people.

In a practical sense the element of the bonus system that workers most frequently encounter is the bogey. Management says it uses the bogey to keep a record of individual performance. Since most workers did not believe that an accurately kept bogey would aid them, they tended to pad their bogies to make them look better. Without padding, efficiency rates would obviously be lower. A worker does not have to be pushed very hard to realize that he or she can look better by using a pencil. Knowing how the bogies tend to be abused, one must be skeptical about the basis of the bonus.

I was surprised at the ingenious ways workers have of beating a rate. No matter how rates are made, workers seem to find more efficient ways to do things. Most workers are willing to share these short cuts with fellow workers, but are reluctant to let the company know about them. This is because there is a general belief that sharing such information with the company will result in pushing up the rates and will not lead to monetary reward for the workers.

There seem to be many inequities to the bonus system. Workers with long service can be in a group with a poor bonus and receive substantially less than a relatively new worker in a high bonus group. Groups seemingly doing similar work can make different bonuses. Of course, some people work better than others and aren't paid individually for what they do.

On top of this, there are psychological pressures that the bonus system can bring forth. People come to work for many different reasons, but most are in the shops to make money. Money is very important to all the people I worked with. When we repeatedly received what was perceived as a low bonus, people began to turn on each other.

After saying all this, I believe that a group bonus could theoretically benefit both the company and the worker. First of all, people produce more with the enticement of extra pay. There is a special feeling at the end of the month when you get the bonus check. Somehow the money seems like extra money, a reward for good work.

Besides, the group bonus adds to the feeling of group solidarity, and I believe group solidarity cuts down on the alienation people otherwise feel. A system of straight salary or individual pay on a piece rate basis might have economic appeal to some, but it tends to isolate people even more than they are otherwise.

There can be little doubt that the current bonus system operated by management is effective in coaxing extra productivity. However, there also can be little doubt that, if the workers wanted to, they could produce more. In order for the bonus system to generate optimum productivity there must be a belief on the part of the workers in the integrity of that system. At this point that belief does not seem to exist. Instead, the workers I spoke with feel that the bonus system is a manipulative device employed by management to procure extra production without fairly paying for that extra production. As long as this general attitude exists, workers will resist the system.

[illegible]

[illegible]

[illegible]

[illegible]

[illegible]

[illegible]

[illegible]

[illegible]

10. THE IMPORTANCE OF FEELING IMPORTANT

"Me, who am I?" asked Mary Bellini.[1] "You know, I'm not really very much of a person. In a big organization like this I'm just a little fish in a big sea. But I think I'll always remember the retirement dinner the girls gave me. It just boosted my ego, it made me feel special. It made me look special in my children's eyes. I'll never forget that my children came and thought I was really somebody."

Mary Bellini is like most of us in her desire to feel special. She would like to find some value in what she does.

The management at Merrimack Valley recognizes the necessity to make people feel important. Beginning with the orientation program, the company attempts to give workers a sense of personal importance. At the same time that new workers are told, "We produce more than one million dollars' worth of parts a day," they are assured that, "Even though there are ten thousand employees at the Valley, the work, the input, and the attendance of each of you is vital to the successful operation of the organization."

Management tries to reduce a worker's sense of the company's bigness by organizing the more than six thousand shop employees into piecework groups, which vary in size from twenty-five to two hundred employees. In order to further foster a group feeling, a worker's bonus is determined by his group's performance.

Large portions of the plant's monthly newsletter are devoted to stories that emphasize the important roles individual shop employees play in the successful operation of the plant. Many supervisors take an interest in worker's lives. June Warren told me, "You're the only person I've ever told this to, but I've told Jack (my supervisor) about my problems with my husband."

The company tries to show its human concern by operating programs to help people with drinking and/or drug problems and by helping individuals with personal problems through unpublicized efforts. Yet, many workers complain about the company's size and impersonality.

Janie Silver told me, "You're just a number here. If you get hurt they won't ask you your name, just your number." Mary Kennedy

[1] The names in this chapter have been changed.

echoed this feeling when she said, "They don't even know you're alive. They want you to show your card at the gate in the morning. You punch a card with a number, and it's your number and not you who gets paid at the end of the week."

Larry Lucchino, a man with twenty-three years of service, who worked previously in the old Haverhill shops, thinks the company has gotten too big. "In here," he said, "they don't care about you, it's too big. In a small shop, a shop of ten or twelve people like where I work at nights, they care about you. You do more work in a shop like that, because they can see the impact of the work you do or don't do. You know what you do makes a difference. You can't feel it here. With all the bullshit they give you about how important you are, you're not part of anything. You're just a number, so you don't care really about how much you do. You do your job and no more."

Carl Zarkowski was always telling me how inefficient the company was. He thought, "This place could operate efficiently, but to do so it would have to have little independent companies within, where we were all dependent not just in word but really dependent upon each other. Then we'd really help each other out. You'd see what would happen then, people would work harder, and they'd enjoy it more. But that will never happen here. The company is too big and that's how they're going to stay."

The company's size and what was viewed as an unresponsive bureaucracy alienated large numbers of workers. However, they were not so alienated that they wouldn't make extra efforts when asked. A little human contact, a little feeling of real, as opposed to programmed, concern could work magic. Time and again I saw people respond to personal requests on the part of supervisors and other personnel.

One Saturday during inventory a worker showed up who had made it clear that he'd never come in on Saturday or work overtime. He had told me previously, "The company owns me for forty hours a week and that's all." I was surprised to see him and asked why he had changed his mind. He said, "Jack asked me to come in. I told him, 'No way.' He told me he really needed me. Well, I wouldn't do it for the company, but when he put it that way I figured I couldn't say no to him."

Personal contact as opposed to impersonal relations often led to very different performances. For example, in our department one of the women, Tracy Harding, used to build models for the department's engineers. She reacted very differently to two engineers.

One came down and told Tracy he wanted her to build something. She told him she was only a 32 grade bench person and didn't know if she could do it. He told her to try. She finally agreed. He said he had some meetings and lunch and would be back that afternoon to pick the model up. She didn't say anything more until he left. Then she said, "He doesn't want to sit here and worry about the board, OK, but he'll come down this afternoon and he'll find I haven't been able to figure it out. If he wants to take two hour lunches that's his business, if he wants me to do him a favor that's my business."

The engineer came back that afternoon to find that the work hadn't been finished. He was upset, but there wasn't anything he could do about it. Tracy stalled for two days, as he came and went, before completing a job she admitted to me later she could have done in a few hours.

There was another engineer who worked with Tracy. He, too, occasionally had models for her to build. He would stay with her for a while, they'd talk about each other's families. He'd ask if there was anything he could get her to make doing the board easier. When he left she told me, "He's a nice boy, he really cares about my helping him out. He explains to me what something is. He doesn't throw it at me and tell me to do it. I try to help him out."

Tracy not only did the work he requested, she suggested a change in the design of one board which she felt would make the work easier for the bench worker.

Still another example of how people respond to personal rather than institutional requests occurred when I worked upstairs on the night shift. A hose leading to a chemical vat broke, causing a large, potentially dangerous spill that spread behind our area and began leaking down to the first floor. It was eight o'clock. The union contract stated that we didn't have to clean it up; it was the responsibiliy of the maintenance crew. We could have left the spill, but our supervisor asked if a couple of guys would help out and clean it up.

There was no hesitation, no statement about that not being our job. Four of us went out back with paper towels and buckets to mop up. One of the fellows went to get some sand to absorb the spill. Wearing face masks to protect us against fumes, we worked steadily for two hours. When we were finished our supervisor took us to the cafeteria for some refreshments. He thanked each of us and told us he knew we didn't have to do what we had done, but he had appreciated it. We weren't rushed through our coffee and doughnuts. When we had finished, we were sent down to the company clinic to make sure we were all right. Any of us could have told the woman in the clinic that we felt sick and we wouldn't have had to go back to work for the rest of the evening and possibly the next evening. Everyone went back to work.

Once back on the job, everyone returned to the normal bitching and moaning about this or that. There were the usual complaints about the way the work was organized. Somehow no connection was made between what we had done a few hours before and what we were doing now. What we had responded to was a concrete request for assistance. The company wasn't asking for assistance in some abstract manner. Our supervisor was in a jam, and he asked for our help and we gave it to him.

In all these examples there is an obvious lesson to be learned. People want to feel needed and necessary, and they respond, almost regardless of the task, when they are made to feel that what they do matters. There is nothing particularly startling about this conclusion. Fifty years ago Western Electric was the site of the famous Hawthorne studies where experiments were conducted to see how changes in the color of walls and intensity of light would affect productivity. No matter how they changed the environment production increased. They concluded that so long as an interest was shown in workers productivity increased. This principle has come to be known as the "Hawthorne effect."

It was clear from my limited experience at the Merrimack Valley Works that some top officials do care about the people working in the shops. Unfortunately, even their good intentions are sometimes sidetracked by the large-scale bureaucracy. Just recently, after signing a nationwide contract with the CWA (Communications Workers of America), Western Electric was struck at many of its plants by the IBEW (International Brotherhood of Electrical Workers). The Merrimack Valley Works, which is 100 per cent CWA organized, was not directly affected, though many other Western Electric plants around the country are IBEW organized and were shut down.

As the strike stretched into the second week, the Merrimack Valley Works felt the impact because stocks of supplies from other plants were seriously lowered. Insufficient supplies were coming in to maintain high level production schedules. The general manager of the Merrimack Valley Works, and other top plant officials made a decision to try to minimize the furloughing of employees and to keep the plant operating effectively. The plant's management was determined to try to ride out the strike and to keep its workers productively employed.

This decision was generally appreciated on the shop floor. As the weeks stretched on and the strike at the other plants was not settled, this plan was jeopardized. Local officials decided that something would have to be done. Again trying to avoid layoffs, the officials came up with a plan that would have had some of the people in the shops take off some of their vacation time. The top officials felt that if enough people voluntarily took some vacation time no layoffs would be necessary.

This all showed a concern for the workers as well as for the company's financial position. However, in telling supervisors of the plan something went wrong. The plan was misunderstood. Several supervisors went out into their groups and *told* rather than asked workers to take days off.

For days after workers were furious about being told to take vacation days. A wildcat strike was threatened. As it turned out the strike with the IBEW was settled, people didn't have to take vacation days, and things returned to normal. However, this unusual occurrence is a perfect example of how a real concern by upper management had somehow been distorted by the bureaucracy. What had been weeks of efforts to maintain the existing work force and good worker-management relations was disastrously undermined. Company officials still believe, and I agree, that, had people been told what the company was trying to do and asked to help by taking days off, the response would have been positive.

Although after a time these incidents fade, such things are never totally forgotten. That is why I didn't find it uncommon to talk to a person who had generally positive feelings about this job and the company while at the same time had some deep-seated resentment because of a past experience.

Jane Hartwig, a person who generally spoke very highly about Western Electric, quite unexpectedly told me of a long-standing grievance that had occurred nearly fifteen years ago.

"My little girl was sick," Jane said. "I stayed home with her, and I called the supervisor to say that I was staying home because my daughter was sick. The next day when I came in, my supervisor sat me down and said. 'You know, Jane, that's not a good excuse, because your daughter is sick is no reason for you not to come in. You're working for a living and we expect you to come in.'

"I couldn't believe he was saying that. I thought maybe I shouldn't say anything but he got me mad, and I told him, 'Now wait right there –my job's important to me and I expect to be paid for what I do, but this little girl comes first. My children are more important to me than my job. If my daughter or son is sick, and if I think they need me, I'm going to take care of them. Now I'm never going to call you again and tell you that one of my children is sick. If they're ever sick again, what I'm going to tell you is that I'm sick.'

"He didn't say anything and ever since then that's what I've done. If one of the kids was sick I'd call up and say I had a virus. I don't like to tell lies, not even white lies, but if the company forces me to do it, that's what I'm going to do."

Later that day I overheard another woman say that she didn't have any work to do and she was thinking of taking a PX (excused absence) the next day. Another woman told her, "Why take a PX? It's not your fault that they don't have work. If the situation was reversed they wouldn't be helping you out, so don't be silly."

After we had received our paychecks for a month when our bonus had fallen, Meg Shepp told me, "You know, in the thirteen years I've worked here I've never understood it–every time they pat you on the fanny, they kick you in the teeth. They just make you feel there's no reason to work hard, there's no reason why you should care. They play with you in a very bad place, they play with you in your pocketbook."

I asked Theano Nikitas, a woman I often talked to, how she felt about the way people were treated. She told me, "You talk to any person who's been here long enough and they've been screwed in some way by the company. Even me, I like this company, it's been good to me, but," she emotionally continued, "they've hurt me too. I was working upstairs. I was on a job where we were supposed to make one hundred boards a day, and we weren't doing more than forty. Well, one day I accidently dropped a piece I had been working on into the solder pool. When I got it out I noticed that it was beautifully soldered, the way it was supposed to be. So I experimented with the next piece and I soon discovered that I could do the board much faster by dropping these pieces in the solder bath. Soon instead of making a hundred units a day I could make two hundred. I showed the four other girls how to do it, and after that we showed the four on the night shift. We were all exceeding the rate and one day I noticed a time study man hanging around. I told him he wasn't supposed to be here, as long as we followed the regular procedure.

"He told me he knew that we were all beating the rate, and he said, 'If you've figured a way to do the job better, why don't you write it up? You could make some money from the idea.' So I wrote it up and I got a note from the company saying they appreciated my suggestion, but they couldn't give me anything because the change I had suggested had already been in the works. In the works, I said, those are my ideas in the works. So I took it up for arbitration. I took it all the way up to the national level. I won and the company gave me $137. I figured from my suggestion they went down from the twelve people working on the boards to two and had increased productivity, and they gave me $137 for that. After that I decided that if I ever found a quicker way to do something I would never give them the suggestion."

I heard this type of complaint more than once. Someone would make a suggestion and would feel cheated. Nevertheless large numbers of workers continue to make suggestions. In the last five years the company has paid out more than ninety thousand dollars to employees for their suggestions. Ron Clay told me he had received a thousand dollars earlier in the year for a suggestion he had made about using a less expensive tape on a job. He told me, "The award came at just the right time to pay off some medical bills. I could have used more. Right now I'm trying to get another suggestion through."

An unresponsive bureaucracy was not the only thing that workers viewed as robbing them of their sense of importance. One of the other things people most frequently talked about was the feeling that they were often made to do work that was unnecessary. They were, they said, given busywork when there was nothing else to do. The company seemed to operate under the premise that idle hands meant mischief.

I heard a good deal about busywork, but I didn't really understand how angry busywork could make you feel until one day in July. Steve, a layout, had asked me to try to finish some boards he was anxious to ship out. I hadn't thought I could complete the job before the day was over. However, I surprised myself and finished the boards a little after the lunch break. The layout was thankful and I asked Jack, our supervisor, what he wanted me to do next.

Jack looked around and then took some boards and some terminals and told me to insert the terminals in the boards. When I began working on the boards, a couple of women stopped by and asked me what I was doing. I told them and they said, "Why is Jack making you hand-insert those parts? Rosie can do them about ten times quicker on the machine."

I kept working, but the next time Jack came by I asked him was it true, could Rosie do them ten times faster on the machine. "Yes," he said, "but she occasionally breaks one of the terminals, so this way even if you build them slower, you won't break any parts." Soon after Jack left Julie came by and asked me why I was hand-inserting the parts. "Because Jack told me to," I said. "Well," she said, "he's just making busywork for you." I told her Jack's explanation. She laughed and said, "He doesn't have anything for you to do, and rather than having you look like you're not doing anything, he's making you hand-insert a board that you could do with a machine."

As I sat there the rest of the day, I really became furious for the first time I could remember. I was doing something that there was no real reason for me to do simply because there wasn't anything else to be done. I worked on those boards for the rest of the afternoon and the entire next day.

The following day of busywork was nearly unbearable. I kept thinking that this was my reward for having finished my work ahead of schedule. For the first time I decided there was no reason I should rush on a job. If I did finish and there weren't new boards, I would be kept busy, regardless of whether the job was necessary. I told one of the women how insulting I had found the busywork and how much it had upset me.

"Now you know how it feels," she said. "It is insulting. Almost all of us who have worked here know that feeling. It's degrading to know that you're being given work just to keep your hands busy. It's hard to feel that you're important when you know that. They let you know that you're expendable. They let you know that no chair and no job is yours. They move you around at will. That's why workers feel unimportant."

The sense many shop people have of feeling unimportant is further buttressed by a double standard. The shop people know that the world of work at the Valley is different for them and the office people. "You know, they brag about this place," Penny said, "about all the windows, and what a nice view it has. Did you ever see a window down here? We don't have the view. The offices have the view and we get to work with no windows."

The windows are just one of the discrepancies. Air conditioning, just put into all shops this winter, was in the offices long ago. One worker told me, "It's the hypocrisy that gets to you, the double standard. They try to sell you this crap about how you're important, and then you find it's only the people in the shop who have to punch in and punch out. We can't smoke, we can't do this or that, but they can. The engineers and the office people have it soft compared to us."

Tom Medichie, a toolmaker with more than twenty-five years of service, told me, "It burns me up that I can't smoke in my shop. I have to sneak into the john for a smoke, and this young girl who's been here less than six months smokes in the office right across the aisle. It just isn't right. I'll be honest with you, it pisses me off."

The work routine of the bench hand is much less democratic than that of the office staff and the engineers. The shop people cannot organize their days or decide which way to do things. Their days aren't organized by them, but for them. Rarely were we told how much we were expected to do in a day. Instead we were given a job. Of course, people do decide how to pace themselves informally. We did not work under the gun of an assembly line, but there was a clearly perceived difference between the work flow in the shop and in the office.

The double standard is put to shopworkers in other ways as well. We were frequently told that the Japanese were giving Western Electric a lot of competition, and that because of this competition we, the shopworkers, would have to work extra hard. One worker was incensed each time this message was given. Finally he told me, "How come the Japanese are only competing with us in the shops? How come we don't hear that they are paying their engineers and managers about half of what these guys get here? How come you never hear that, just that they're competing with us in the shops?"

Not all of the reasons why people feel their work is unimportant can

be laid on the doorstep of the Western Electric plant. The value system of our society relegates factory work to a relatively low status. At the same time society claims all work is important, in many different ways it makes the judgment that factory work is not only unimportant, but that factory jobs are occupied by people who have no other skills. In fact, we refer to much of the work done by large numbers of people who work in plants like Western Electric as unskilled labor.

Many of the workers I met know that the labor market for them is limited. However, they maintain the illusion of a different situation in one of two ways. Many talk about not having their children end up doing the work they do, just as many of their parents talked about them not working in the textile mills or shoeshops as they had. Others talk about someday leaving the shops. I heard workers with as little as six months' experience, and as much as twenty-five years, talk about leaving the company and taking other, more interesting, challenging jobs. Relatively few of them will leave Western Electric.

I was surprised to discover how many people I met wanted to do a good job and cared about what they produced. Sure people bitched, told me they hated their work, told me they didn't care, but I saw that they did care. No matter how small the job they were asked to do, they tried to do it well. There were, of course, some people who really didn't care, but the intricate sets of rules and the constant feeling of being watched didn't make them good or even better workers.

Much has been written about the lack of pride shown by factory workers in their work, but from what I could see workers do care about what they produce. What alienates them is the structure of the rules, the undemocratic nature of the work situation, and the gnawing feeling that they are unimportant. Work is all too frequently organized in a way that makes people feel that they are not expected to be responsible, that their contributions are unimportant, and that they are interchangeable with machines and with each other.

The company's efforts to minimize the effects of plant size have revolved primarily around the small work groups. However, these groups are artificial. They do not decide rules, production techniques, ways of doing business, handling bogies or operating procedure. Everyone follows plant-wide procedures.

Occasionally you can see how people would like to feel about their work. In 1973 top management gave the shop people the opportunity to run an open house. Shop and plant-wide committees were set up to organize it. All of the arrangements were made by workers, and the Works had a three-day open house during which workers could bring families and friends into the plant. The open house was a huge, unqualified success. People roamed through the plant, showing their children what it was they did. One could feel the pride people took in showing their families their work, in making their jobs sound important.

The work is important and management realizes it is important. Unfortunately much of this appreciation is not felt on a regular continuing basis out on the shop floor. The long service records of a majority of employees are witness to the fact that people find reasons to stay at the Works. In fact, most of the people I met liked the company and thought it was better than most in the area. Nevertheless large numbers had long-standing grievances and felt inconsequential in their work They have learned to internalize these grievances and feelings.

TO: ALL PERSONNEL

SUBJECT: EARLY RETIREMENT PROGRAM

AS A RESULT OF AUTOMATION AS WELL AS A DECLINING WORK LOAD, MANAGEMENT MUST OF NECESSITY, TAKE STEPS TO REDUCE OUR WORK FORCE.

A REDUCTION IN STAFF PLAN HAS BEEN DEVELOPED WHICH APPEARS TO BE THE MOST EQUITABLE UNDER THE CIRCUMSTANCES.

UNDER THE PLAN, OLDER EMPLOYEES WILL BE PLACED ON EARLY RETIREMENT, THUS PERMITTING THE RETENTION OF THOSE EMPLOYEES WHO REPRESENT THE FUTURE OF THE COMPANY.

THEREFORE, A PROGRAM TO PHASE OUT OLDER PERSONNEL BY THE END OF THE CURRENT FISCAL YEAR VIA EARLY RETIREMENT WILL BE PLACED INTO EFFECT IMMEDIATELY. THE PROGRAM SHALL BE KNOWN AS "RAPE" (RETIRE AGED PERSONNEL EARLY).

EMPLOYEES WHO ARE "RAPED" WILL BE GIVEN THE OPPORTUNITY TO SEEK OTHER JOBS WITHIN THE COMPANY, PROVIDED THAT WHILE THEY ARE BEING RAPED, THEY REQUEST A REVIEW OF THEIR EMPLOYMENT STATUS BEFORE ACTUAL RETIREMENT TAKES PLACE. THIS PHASE OF THE OPERATION IS CALLED "SCREW" (SURVEY OF CAPABILITIES OF RETIRED EARLY WORKERS).

ALL EMPLOYEES WHO HAVE BEEN "RAPED" AND "SCREWED" MAY THEN APPLY FOR A FINAL REVIEW.

THIS WILL BE CALLED "SHAFT" (STUDY BY HIGHER AUTHORITY FOLLOWING TERMINATION).

PROGRAM POLICY DICTATES THAT EMPLOYEES MAY BE "RAPED" ONCE AND "SCREWED" TWICE, BUT MAY GET THE "SHAFT" AS MANY TIMES AS THE COMPANY DEEMS APPROPRIATE.

11. A FARM FAMILY

Greta Barnes—Greta Barnes was one of three process checkers in our shop. She would come by two or three times a day to sample completed work and to check for mistakes. There was "no charge" for the conversation and philosophy that accompanied the daily checks. I looked forward to our daily conversations, which covered a lot of ground.

It was five o'clock when I walked through the Barneses' front door, past two pairs of muddy boots. A wood fire was burning in the old Crescent Queen stove that dominates the kitchen. Greta Barnes and her mother, "Granny," were busy preparing dinner and complaining about the day's spring cleaning. The farmhouse is casual, somewhat untidy, and very livable. The furniture is comfortable and functional. The couches retain the shapes of visitors long after they depart. Spring cleaning is marked by cursing, laughing, and complaining. Wayne and Lynn, Greta's daughters, have stayed home from work and school to help. They have cleaned the entire house and stripped the linoleum from the dining-room floor, laying bare the wood. Over and over they rehash the day's cleaning. My presence is hardly noticed.

Finally, everyone is called to dinner. Buddy, Greta's husband, is home from work, and Arthur, her son, is just up from a nap. Only Richard, the youngest child, is absent, off at Little League. Dinner will have to be saved for him.

The table is covered with pork, potatoes, corn, and applesauce—all grown on the farm. Hands begin to reach, and food is passed around. Potatoes missed the first time around will pass your way more than once again. Dinner is a noisy time. Stories and jokes dominate the conversation. The day's events are shared while the vast pile of food on the table is whittled away. Story falls upon story, and the food keeps getting passed. A plate once emptied is filled again.

The women clear the table and wash the dishes. After the dishes Granny, retiring to her favorite chair in a corner of an alcove off the kitchen, begins to read a book. Greta and Lynn turn the dining room into a fitting and pattern-cutting room for a new dress. They shout, yell, and laugh as they attempt to work together on Lynn's dress—each has a better way to do it.

Wayne, who has temporarily disappeared upstairs, reappears to claim Lynn for choir practice, pulling her off a chair as she's being fitted. Greta, left with the pattern, watches her daughters walk to the door and calls after them in mock anger, "You rotten kids."

Buddy and Arthur have put on their weather-worn coats, small woolen caps, and crusty boots and have headed to the barn for the spring night chores.

Greta and Buddy Barnes had a dream when they got married. They wanted to own their own farm and they wanted their children to grow up on it. In 1949, two years after they got married, they bought the farm they still live on. Their children, Arthur (twenty-two), Wayne (nineteen), Lynn (seventeen), and Richard (thirteen), have lived there all their lives.

"When we bought this place," Buddy says, "there was the house and fifty-five acres. We paid twelve thousand dollars. We didn't have any money, to speak of, so we borrowed some from my sister, and a thousand dollars from the Beneficial Loan Company for the down payment."

In order to help meet the mortgage payments, the Barneses began raising poultry for a Mr. Spencer. "It was a real pain," Greta says.

"The deal was this," says Buddy. "Mr. Spencer would give you all the birds you were capable of handling, and pay you a cent and a half a bird per week. He'd supply heat and everything the birds ate. We'd supply the labor—that's what the cent and a half was for.

"At first Mr. Spencer was rather reluctant about giving us chickens because we had about fifty chickens and a couple of cows of our own. Right off the bat he was suspicious about whether we were going to use the grain for our animals. He didn't come right out and say it, but I told him that if he thought we were going to use his medicated grain to feed our animals, he could take his damn birds off the farm.

"His birds were being forced to eat so much they had to use the medication. We couldn't see it. Ever since I was a kid, I always believed in organic gardening. Even now we don't use fertilizers or poisonous sprays on the garden. I don't know whether the kids told you, but one of their duties in the summer when we grow potatoes is to go out and pick the potato bugs off the potato plants. They take a can of oil and every night go out to the potato patch and pick off the potato bugs.

"I don't rightly know how I got off on that, but with the chickens it was a lot more work than we figured."

"It was a happy day," Greta says, "when we told Mr. Spencer we had had enough of the chickens."

"When we bought the farm," Buddy said, "I'd have liked just to farm, but the place was stripped of all machinery—stripped of everything as far as workings. Naturally, I had to buy a tractor and haymaking equipment, a plow and harrows, and what-have-you. So I had to continue to work for a living. It meant doing the farm work as a second job."

"I went to work," says Greta, "because we needed the money. When we first got here, Buddy was making two bucks an hour. We had debts, almost no furniture, and we had to borrow the down payment for the house. Besides, with my mother living with us, it wouldn't have worked out well if we both stayed home. My mother and I have a clash of personalities, just like Wayne and me. I realized that it was wiser for me to be out working, and my mother to be taking care of the kids. Six weeks after I had my second child, I began working at Western Electric.

"If my mother hadn't been here to help raise the kids, Buddy and I would have had to work different shifts. That would have meant we'd hardly get to see each other."

All four children feel close to their grandmother, who has been, they all say, like a third parent. She is the one all the kids remember waiting at the door when they returned from school. Richard says, "Some of the kids at school talk about coming home to an empty house, but there was always Granny here."

Lynn says, "I love her. I mean she didn't have to live with us, and raise us kids, but she did. I'm really glad she did. All my friends love her too. They can't help but call her 'Granny.' She's wonderful, and she's practical, too. She'll tell you, when you're daydreaming or something, 'Spit in one hand and wish in the other and see which weighs more.' "

Recently Granny had a stroke, and it upset everyone in the family. Richard, of all the kids, took it the hardest. During dinner on the Monday following her stroke, Richard asked, "What happens if Grandma should get sick during the night?"

His grandmother answered, "Well, if I get sick . . . we all have to die sometime or another."

"We didn't talk about it anymore," Greta says, "but Richard vomited Monday night. He didn't go to school on Tuesday because he told me he wanted to stay home with Granny just to make sure she'd be OK. I told him that would be OK and if he wanted to stay home the next day, that was all right too."

Granny has recovered. She is a strong, stubborn woman, like her daughter Greta.

"Having my mother live with us hasn't been easy," Greta says. "I think it's been hardest on Buddy, but my mother means so much to all of us. There are tensions when any two generations of adults live under the same roof, but we've pretty well learned to live with them.

"A family has got to make a lot of adjustments, that's life," Greta continues. "I'll tell you, having both parents work, like we do, changes things for everyone in the family. For a woman it really means holding two jobs: a work job and taking care of the home. My house is always a mess and it doesn't bother any of us, but I can't imagine how the women at work who need to keep their places spotless do it.

"I think a woman needing to work is also tough on the man. He gives up a lot. With his wife working, he gives up the American sense of dominance, of 'I can provide.' Really, the whole family gives up. A teen-age daughter may have to do more cooking than she would in another family. Everybody has to pull together in a different way when both parents work.

"Buddy and I decided the only way to do it was to limit our jobs to forty hours. I never do overtime, or work on the weekends. Sometimes Buddy does put in a few hours in overtime. Although that's where the best money is to be made, we figure we'd like to spend as much time as we can with our family."

Both Buddy and Greta are diligent workers, but neither is really satisfied by factory work. Greta says, "I think I'm a good worker, but do I like my job? Not particularly. To be honest, I think I have a lot to contribute. I don't think I get a chance to do it at work. I suppose that's one reason I'm so involved in activities around where we live. Like the Grange that we belong to needed to raise money. So I got a hold of my mother, and we did up six turkeys from our farm for a turkey dinner. We peeled the potatoes and made the meal for 175 people. We got fiddlers and arranged the whole evening. There was dancing and fiddling. Everyone had a grand time. The Grange makes me feel important, like I can really contribute something."

Buddy doesn't complain about work, but then Buddy doesn't strike one as complaining about anything. Arthur says that his father's original intention when he bought the farm was to farm, but that he didn't have the money to start and had to continue his job. "Everything he does," Arthur says, "makes you know how much he enjoys farming. You give him a patch of potatoes and a hoe and he's the happiest guy around."

Buddy puts in a lot of extra hours on the farm after work, and on the weekends, but he says, "I don't mind it because on the farm you're not working under pressure. When you're working in the shop, they expect you to do a certain amount of work regardless, even though they don't stand there and say you gotta do so much. Still, they push you to a degree. You try to give them a day's work for the money they give you. But when you're home, if it takes you an hour to do it, OK. If it takes two hours, there's nobody there saying you got to have it done in the required time. The guys down at the shop don't understand why I put so much time into the farm. They think I work too hard. What they don't understand is that for me the farm work is relaxing."

PLUS
25
6014
EVEREADY

Over the years, the farm has shrunk from fifty-five acres to thirty-five acres. Twice the Barneses have had to sell pieces of land to pay bills.

"When Lynn was only eighteen months old," says Greta, "she needed a heart operation. A few years later Wayne got real sick with infected kidneys."

"One week to the day later," Buddy said, "Arthur came down with the same virus. Three weeks later, Lynn caught it. We hadn't fully recovered financially from the expenses of Lynn's operation, and all of a sudden we had three kids in the hospital.

"Blue Cross took care of some of it, but we had to take care of a lot of expenses. There were doctor bills and penicillin. We had to put central heating in the house before the doctor would let the kids come home.

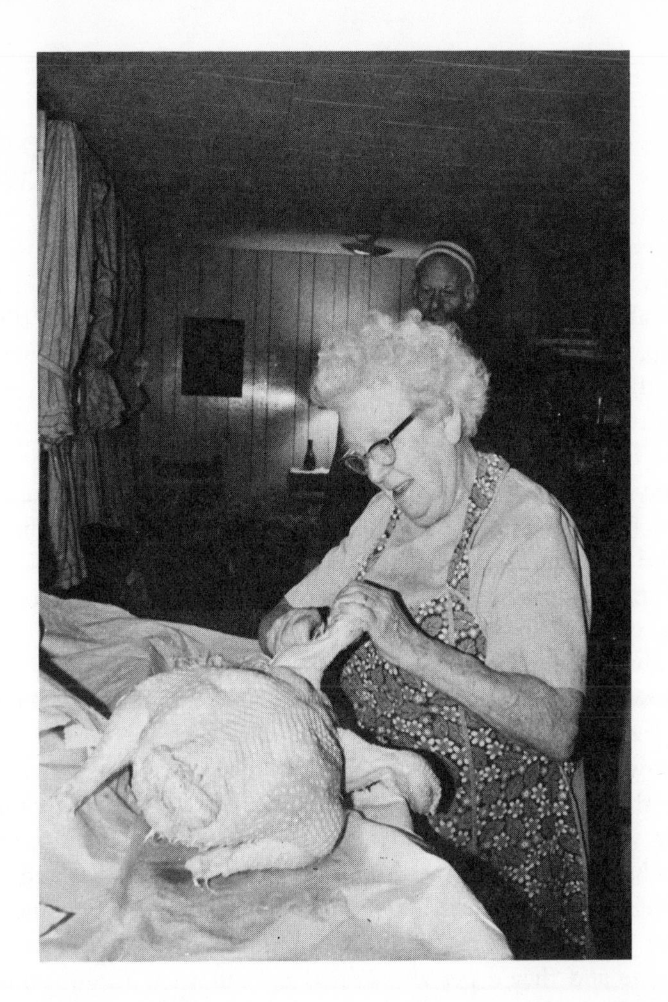

"It was getting to the point where we were pretty embarrassed financially. We couldn't see how we could bail ourselves out. We didn't want to take out any more loans because we couldn't see how we could pay them back. So, we sold ten acres of land and paid off all the bills. We sold another ten acres a couple of years later."

"We've had a lot of offers to sell more of the land," Greta says. "All these developers would like to buy the land. We've been offered a lot of money for our remaining thirty-five acres, more money than we could ever have otherwise. But," she says, "what would we do with the money? We have the kind of place that we always wanted. Why should we want to move? Lots of people leave the countryside to go to the city to find work. They work all their lives to make enough so that they can buy some land in the country to retire on. We've had it all along, so I guess we're lucky."

At the Barneses, leisure, as well as work, is a family activity. "I don't think we ever really talked about it that much," says Greta. "Our own parents were that way, spending most of their time with us. We just naturally wanted to be that way with the kids."

"We've always tried to stress one thing," Buddy says, "and that is if the kids worked, they deserved to be provided with recreation. There's always been plenty of work for all of the kids. When the girls grew up, they had to take and do their share of the outside work as well as the inside work. As far as I'm concerned, there's been a good many girls who were raised on a farm who were out taking care of animals. There was no problem with their growing up, so why should it be any different with my children? And, naturally it taught the children the responsibility of taking care of the animals in the event that anything happened to me or Greta and we couldn't do it.

"Naturally, if they've got to do that work, you've got to show some kind of recreation for them, and get them all involved in it. That's the way I always felt about it. When we first came here, there was no pond down back. And the summers used to be pretty hot. The kids would want to go swimming. So we used to take them to Harrison's pond down the road. It wasn't too bad when we first moved here because they only used to charge ten cents a person to go swimming. Then it started going up to twenty-five cents, fifty cents, a dollar. Well, when we came home from work and the kids wanted to go swimming, we'd take them down swimming. Then we'd come back, and after we had supper we'd do the barn work or whatever else had to be done and they'd get all sweated up and want to go swimming again. Well, back to Harrison's. It began to be a pretty expensive proposition. Then we had a dry spell, I had a pond dug out down back, so that solved the swimming problem. Whenever the kids came home and it was warm enough for them to go swimming, they just went swimming. So, naturally, it made a place for them to go skating in the winter. If there was no skating, we'd have beano games and card games. At Christmas time, we used to get all kinds of games that we could play together."

Wayne says, "We've always been a close family, doing things together. We've always had our own little community. My parents didn't seem to go out a lot by themselves or with friends on Friday or Saturday nights like a lot of other parents.

"The only time they went out a lot was when they went square dancing. They'd go out square dancing, and we kids would stay home with Granny. But then when we got old enough, they taught us and had us take square-dancing lessons. When we got old enough to join the club, we'd go square dancing with them.

"They used to take us to the movies when we were little. Or what I used to love was the bingo games we had. Saturday would come and we'd go to the candy store and buy all kinds of penny candy. Then we'd come home and wait for the bingo game. Every time you won a bingo game you got a piece of candy. We'd play for hours–straight games, *h*'s, crosses–fill up the whole board. My father never played. He was always the caller. When someone won, he got candy too."

Greta and Buddy smile when they hear about Wayne's memory. Greta says, "The kids did all sorts of things. We were very close with their allowances. None of them got more than sixty cents a week. So, to earn extra money, they used to put on plays. They'd rehearse a play for months and then give a show. "They'd sell popcorn at ten cents a bag. The plays were pretty good, too."

"They'd put on circuses, too," says Buddy. "They'd do horseback riding and have a regular show."

Greta says, "We used to have skating parties. We had swimming parties. We had hayrides for them."

"Even when they got older," Buddy says, "we used to think of them when we thought of recreation. It was never a question of just Greta and I going on a trip and leaving the kids home. We'd always plan on taking the whole family. It was more fun that way."

Both Lynn and Wayne consider each other their respective best friends. "I look at some of my friends," Lynn says, "and they can't stand their sisters. They hate their sisters. They wish they'd move out or something. But I don't want that to happen with Wayne and me. We've always been close. We became friends because there aren't so many people out here. We have stayed close because, well, we love each other."

Wayne says, "Lynn, aside from being my sister, is my best friend. I'm the one that always gets myself into embarrassing situations and she's the one that kind of comes along and smooths it over. Lynn and I do a lot together, and we use 'the money.' It's always been 'the money.' I get paid on Saturday and so I take us through the weekend. Lynn gets paid on Monday and so she takes us through the week. I make more so naturally I kick in more."

Like the two girls, the two boys have a very special relationship. Although there are nine years' difference between them, they are very close. Richard and Arthur share a room. To make some extra spending money over the summer, Richard rode on Arthur's milk route. Arthur paid him a dollar a day and bought his lunch.

"I think we're all very close," Arthur says. "The only thing is as we get older we don't say 'I love you' so often to each other. I don't know, maybe it becomes more embarrassing to say it as you grow older.

"A lot of things change. Things have changed between my dad and me. I don't know if this is the typical stage I'm supposed to be going through—where I'm trying to express my manhood. You know he's the dominant male and I'm competing with him. When we have an argument I voice my disagreement, but I really can't follow through. I do a lot of things under protest, because I think his methods are wrong. Still I have a lot of respect for him.

"He's really strong, almost to the point where he could be called stubborn. He has the courage of his convictions and, though he's never really come out like a religious fanatic, he's got strong moral views. I think there are very few fathers like that in my peer group . . . they all want to be one of the boys.

"He doesn't try to be one of the boys. He's my father. I'll tell you the kind of thing my father would do. He signs his paycheck over to my mother because she takes care of the finances. She gives him five dollars a week. It's really not an allowance, just some spending money. When I was going to U.N.H. [University of New Hampshire], he was sending me two or three bucks out of that so I'd have a little spending money for beer or whatever. That's something more than a buddy would do."

Wayne, like Arthur, is feeling the tension of being partly grown up and yet at the same time partly her parents' child. She says, "I know that no matter what I do, no matter what predicament I get myself in, my family will be there to support me. Sometimes," she says, "I think they're too strict. For example, even though it isn't enforced, I feel like they want me home before midnight. I very rarely come home past twelve. Although they don't say anything about it, I feel their displeasure when I come in late. I'm nearly twenty years old and when I go out, I shouldn't feel bad if I come in late.

"Sometimes I question the authority, which you don't do to my father. My father was brought up in the old school, where the father is boss. I just can't buy it. I feel that he can be wrong. We don't have calm discussions because the minute I start disagreeing at all he starts the 'I am the father' act. I boil with anger and I start yelling, and he starts yelling. But then it's over.

"My mother and I are both headstrong, and this sometimes causes problems between us. Sometimes I swear I'll never speak to her again. Once I didn't speak to her for two days. It was terrible. I was glad when it was over and by then I didn't even remember what made me so mad. With all the fights we've had, I wouldn't choose another mother. For all the times she's called me a bitch, there have been more times when she says to me, 'Have I told you I love you?' It really makes you feel good."

Neither Buddy nor Greta has tried to push the children into farming or anything else. Greta says she wants her children to be happy. "I don't want them to get ulcers," she says. Buddy says, "I don't care what they do just so they try to be good at whatever it is they do."

There has for a long time been talk among the children of building homes for their families in a section of the farm called the pine grove. It is both a dream, and a very possible reality. The ties to the farm are strong.

"This farm means a lot to all of us," Arthur says. "For me there's a feeling of accomplishment you can get on a farm which you don't get elsewhere. Like when I sit down at Thanksgiving dinner; 90 per cent of the food on the table came from the farm, and we all had a hand on it. Like my mother says, 'Seeing food you've grown on the table is more rewarding than bringing home a paycheck.'

"I think I'd like to make to make a go of farming. I'd really like to work the farm. You're your own boss. You're self-employed. It's not a hassle. You may have to work ten hours a day, but if you can make a buck working for yourself it's worth five dollars working for someone else.

"That's what I'd like to do, but I don't know if I really have enough to try it. It's always the security. Somewhere's along the line I developed the attitude that you gotta have security. I would very much like to get a degree in some field like engineering so that, say the farming fell through, I could always fall back on something else."

Wayne is currently working at a funeral parlor. Since graduating from high school nearly two years ago, she has held a variety of jobs. For a while, she thought of joining a religious theater troupe in California, and lately she has thought about going to school for horsemanship. "There are so many things I want to do," Wayne says,

"but it's hard to say what I'll do. I'd like to live close to my family, for all of us have a place in the pine grove, but there are so many things that I want to do it's hardly fair that I won't be able to do them all."

Summer is quickly approaching. The hot weather will soon set in. The heavy farm work will have to be done. It will be much easier this summer than in years past because the family has just purchased a new tractor.

It is a major investment that affects everyone. "See," says Arthur, "we have an old tractor and she's tired. So my father and I went down and looked at the different tractor dealers. We found what we thought was a pretty good buy on this year-old International 415. We needed five hundred dollars for a down payment. Well, we weren't too sure of whether or not we could scrounge it up so I said, 'Have no fear.' I got the five hundred dollars for the down payment from my savings and we got the new tractor. It's already made work easier. We spread manure around six weeks ago which, in past years, would have been impossible. With the old manure spreader, if there was any snow on the ground at all, the wheels would just slide. With this new spreader, it didn't make any difference what it was running on, the tread on the tires dug in and it spread. And the old H can still pull, so just the fact that there are two spreaders cuts your work in half and makes it easier on everybody."

Wayne thinks of the tractor much more in terms of what it means to her father. "It's going to make everybody's life easier, especially my father's. My God, he's fifty-four years old; he does twice the work everybody else does around here. If you could have seen him this

past summer with the baler and the tractor and just being so totally exhausted. Even if buying the tractor means giving up some things it's worth it to me to know that my father will not have to work as hard with this tractor."

The warm weather means the pond will soon be ready for late night swims. Last year on a warm night, Richard and his mother were out in the pond when it started to rain.

Greta said, "We'd better get out of the water, because if the lightning hits the pond we could die."

Richard asked, "I'm in it too. Will I get hurt, Mom?"

"You might and you might not."

"If it hits the water would we both die?"

"Probably."

"If we both die, will our family bury us?"

"Yes."

"In one coffin?"

"Probably not."

"Well, would they give us a discount on the coffins?"

"Probably not."

"It would probably be a very expensive funeral then, wouldn't it?"

"Yup."

"Well," Richard said, "maybe we'd better get out."

12. A SHOP STEWARD SPEAKS

IRENE LAMBERT—I met Irene Lambert six months after I stopped work, and we hit it off immediately. The more I get to know her the better I like her. As is the case with all other shop stewards, she volunteered for the job and then needed to be approved by the executive board of the union. A shop steward's primary responsibility is to handle the grievances of the workers in her shop.

A For Sale sign sits on the tiny brownish-green front lawn at 12 New Hampshire Avenue. Irene Lambert, the owner of the two-family house and occupant of the upstairs duplex, is sitting in her attractive living room talking about being a light sleeper.

"I am usually up at twelve-thirty or one walking around to see if the kids are all right. I like to peek in their rooms and see them sleeping. I know I don't need to look in, but it makes me feel good, and I've done it since they were little. Anyway, I'm a light sleeper. I never reach a sound sleep. I'm one of those people who can take a nap for an hour and I'm feeling good. After supper, I'll often fall asleep for a half hour or so right over there in that stupid chair and get a crick in my neck. If I was to lie down I wouldn't fall asleep.

"Right now my biggest concern is selling the house. What I'm trying to do is to get into this low income housing. I'm sweating it out, because we are just getting by. I earn just enough to support my family. My money's tied up. All this [she points around the house] you see my children and I did. We are four thousand dollars in debt, so it comes out of my pay. After everything has been taken from my check we have about twenty-five dollars a week to live on and we're making it. When the divorce comes through, and the house is sold, we'll pay my bills, and there won't be so much pressure on me.

"I resent my husband for making it so tough on us. He had more than that. Why should they have less? They shouldn't have to worry about money and I shouldn't need to work ten hours a day. John and I were married for nearly twenty-three years, and it isn't right. He's taken away their chance to have their full youth, and I hold that against him."

Nathan passes through the living room and Irene introduces us. He quickly moves on. Irene looks after him and says, "He's a fine young man. All my children are good, I'll tell you they're my real strength. This one, he'll be nineteen in November. He's a musician; he's a drummer. He wants to go to Berkley (music school) and I intend to see that he gets there.

"Robin's the oldest (twenty-one), and she's my only daughter. Now she may say I'm out to lunch, but I think she's mixed up, and hurting because of the trouble John and I had. When Robin got out of high school, she got a full scholarship to Salem State. She didn't take it. It almost blew my mind. She wouldn't go. She wanted to party, and that's just what she does, she parties. She's a beautiful girl, and I know she does everything there is to do. One day she came to me and she said she was pregnant. I almost—typical—I almost went into orbit. When I came down from it, I said, 'Well, if you are, you're coming home.' She came home, but she wasn't pregnant. She has her own apartment now.

"The middle boy, John, this is the one that's closest to me. Parents say they don't have favorites, but I think you do. There's a closeness there that's special. I try very hard not to show my feelings but undoubtedly I must. The youngest one, Yancy, he's the baby (fourteen) and he's paid for not having a father.

"I feel very badly about that because I feel boys need men around them. I've been lucky with the friends that I have around here, because the males have taken over. But it's not the same thing as a father, and I resent John for that. We have four beautiful kids and he's not treating them right. I didn't want our split-up to be like everyone else's. I didn't want to be saying, 'He's a no good bum,' and everything like that. I didn't want that but that's how it's turning out.

"When he left me I still loved him. Hell, I only just stopped loving him a year ago. And now there isn't anything he can do but just keep out of my way. The less I see of him the better I like it. Really if I were a man I'd knock him out because he needs a beating.

"He isn't a mean man, you'd like him, but he's treated me and the children wrong. He's doing his own thing, going through those changes men are supposed to go through. Right now he's down in Cambridge being Joe College. This man is forty-three and he's at school taking up dancing. He paints, he goes to plays, he's become a camera bug, and all that shit. So be it, but he shouldn't ignore his children. I know it's a terrible thing to say, but I hope he hurts. I hope he gets his and he will. You just don't sow without reaping and he's sown some pretty damn rotten seeds.

"It wasn't always that way between us. We were married for nearly twenty-three years. My husband was actually only the second guy I ever went out with, and now looking back I realize that wasn't good. I came from a very sheltered background.

"There were only three black families in town [Andover]. My family was very well respected. We certainly weren't wealthy, but I had everything everyone else had. If my mother couldn't buy it, she made it. My mother, who is seventy-eight and the oldest member of her church, was born in Methuen. Her maiden name was Murphy. I guess when her people fled slavery from the South they must have run up to Canada and they settled in an area where the Irish had migrated, and that's how they took the name Murphy. My grandfather on my father's side was brought up here from Kentucky as a groom for a rich man named Searles. So I come from good parentage with solid roots in this area.

"The trouble was there weren't hardly any black kids where I was brought up. I grew up among whites. It was all right to intermingle when I was a little kid, but when I got into my productive years the parents started to be wary. I was a nice kid, but, you know, it was the time when you drew the line. She's black, you're white. It hurt me.

"John wouldn't have been my father and mother's choice for me. He came from a very good family of churchgoers like ours, but John is three shades darker than me. I'm not color conscious, but my father wanted all of us to marry lighter. In fact, he would have liked me to marry the other side. You've got to remember they came from the real old school. They thought things would be easier for us that way.

"I wanted my children to grow up in a different environment than me, and that was why I was glad when John picked Haverhill to live in. There are more black in Haverhill. I wanted them to be around as many blacks as possible.

"I have had to make some adjustments. I grew up as a white, yet I'm black. I had to fight for my place with my people. Some of them still resent me. Everything about me is different from my black brothers and sisters. Most of them are Southern imports, and I don't think like they do. I wish I could and then in some ways I'm glad I don't. I'm not flashy. I don't think it's necessary. I don't like the way many of them talk. I don't want to talk like that. My father had me take elocution lessons. Yet still I feel that I ought to be able to identify with them. I

belong to the local Elks, which is primarily black, and I enjoy going down there and being with my people.

"I have my hangups, yet when it comes to my children intermarrying, I'm as staunch my way as whites are the other way. I don't think there's anything better than a black. I think we enjoy life, that we enjoy the simple things in life more than whites.

"I'm at my sons all the time about it. Little white girls chase my Nathan from morning to night. I try very hard to be polite to my son's girl friend, but I don't think I fool her for one minute. I want her to know just where she stands with me. I do resent her. I wish she were black.

"I stopped working five months after I got married. I didn't go back to work until Yancy was born. My husband didn't want me to work.

But in the same token, and I'm not being nasty, my husband's cheap. If he earns ten bucks, he'll give it to you but he wants to tell you how to spend it. If I needed underwear or stockings he used to make me wait, and I resented it very much. I'm not someone to ask you more than once. If you turn me down after the first time, I won't ask you again. I put up with it for years, but finally I decided to get a job. I wanted to be more independent. This was fifteen years ago.

"I went over to Western and applied for a job as an IBM operator. When I went into the office I had no thought that I wouldn't be hired. I was skilled (IBM course); I was sure of myself. I was Irene Lambert. I figured here I am, you lucky devils.

"There was another girl who applied at the same time and they took her. She had less training than me, but she was white. The time was not right, and I haven't forgotten that. I never have let them forget it either.

"I didn't apply for another job until after Yancy was born and then I went and applied for a job in the plant. I didn't holler discrimination, I was not as I am now; I wouldn't say five words then, I never spoke up. If you had asked me for an interview I would have said no. I minded my own business fifteen years ago.

"My first job was on the second shift. Work got slack and they wanted me to go on days and I couldn't because Yancy was too young and I was working what we call the mother's shift, so that I wouldn't have to leave Yancy alone for too long. So I stayed out of work until there was an opening on the second shift. I was out three weeks to the day and I came back in again and went to work for Don Darnell.[1] This is, as far as I'm concerned, one supervisor that Western should get rid of; he's a bigot. He doesn't come out with it openly, but he's a bigot.

"When I first started out for him everything was great. I was a tremendous worker. There wasn't a job that he put me on that I didn't make over 130 per cent on. Then he put me on this job and I couldn't do it. If I made the rate, I did poor work. If I did good work, I couldn't make the rate. After doing five or six jobs over the percentage he was going to fire me for not doing well on this one job.

"That experience with Don Darnell changed me. I'm very arrogant now. I wasn't that way when I started out. I had to be abused to become that way. When I started there I was just like the rest of the little sheep ready to be slaughtered. This Don Darnell hurt me. I gave him the best I could. That did it, I became a union steward. I had

[1] This person's real name has been changed.

joined the union the day I went into the plant. My father's a union musician. He taught me the history of unions when I was young, so I've always been very union conscious.

"I had such a chip on my shoulder after that. There wasn't a grievance I lost in five straight years. Didn't lose one, and I'm up against men who are trained to best me. The stronger I got, the worse I got. I tell them you had me as a saint and didn't know what to do with me. Now you've got me as a devil and you'll live with it. I'm a product of Western Electric and what they did in that plant.

"I worked hard at being a good steward, and I'm a damn good one. To this day I put a lot into it. When I handle a case I give it my all. I don't short-change the people, but I only grieve cases that I believe in. I won't touch you if I think you're wrong, and I'll tell you that. I know a lot of people don't like me. But they come to me with their grievances because they know that if I take it, I'll fight like hell for them.

"The year after I became a steward, I became an officer on the Executive Board. I remained on the Executive Board up until the time I ran for secretary. I won and was secretary for four years until I was defeated just about two years ago. When I lost it almost destroyed me. I never lost before. I was crying; I was crushed. I'm not a good loser. I do not like to lose.

"I told the people, you failed me. I never failed you, but you failed me. I was going to resign. But Mike Greico [former president of the local], the love of my life, he told me don't resign, stay a steward, the people still need you. So I stayed.

"Sometimes I don't know why I'm a steward. It's a lonely job. They damn you if you do and they damn you if you don't. You can win ten thousand cases but if you lose one, you're a no-good bastard. If the people don't like the contract it's your fault. You're abused by the people, by the company, even the union. But you hang in there. You can't explain it. By rights I should be out of it. I've never received anything for it. No money or anything. The only thing I've received is power, and I enjoy it. I think I enjoy it more because I'm black and I figure I'm up here and these white people are dependent upon me when they've got problems.

"I wouldn't say this is a bad place to work, but there is a lot of phony shit that they do around here that frosts my ass. For example, they spend gobs of money on that damn lawn out front, and then they fight like hell to keep from giving people an extra penny. Worse than that is what they go through when they have a visitor. Whenever they have a visitor they put it on, and you know it's so phony. Sometimes we've got to change our breaks and we're supposed to clean up our area if an "important" visitor is coming through. Hell, if they want to come in let them see me on my break or at work. And cleaning up the area: fuck that. If I can work in the filth they can see it too. They're no better than me.

"As soon as they get rid of that happy horseshit and get down to the business of dealing with the people, that's when they're really going to have something. They need to treat people better. They need to treat you like a human being, not a number. I'm not Irene Lambert in there, I'm number 24505.

"Everyone knows they have to work, but don't make people come in with the feeling that, ugh, here I am again. Let them come in feeling, OK, I've got to work, but I've got my music and I can relax and do my job and this is coming about at least in my department. It's not a big thing, but it's a beginning, they've never had it before, and I feel if they can go along in these veins, it will change things. That's why I tell people stand behind John [department chief] because he's sticking his neck out trying to make things better, and a lot of people would like to see these experiments fail.

"I'm not quiet about how I feel. I let everyone know. When the company shrink came, I told him, 'Get rid of these goddamn green walls. Give these people a little color. Let them come in and see something besides canary yellow or bedpan blue. Give the women a lounge where, if they don't feel too chipper, they can go and hold their heads for a few minutes. Give them the things that make them feel like a human being. There are always those that will take advantage of this and run it into the ground but percentagewise give them the benefit of the doubt and give them a chance. Let them fail before you say they failed.

"Sometimes they forget that we're Western Electric. I tell them you're nothing without us. Stop treating us like we're second-class citizens because when the shit comes down and we decide we want to get the hell out of here, there isn't going to be a plant. You can have a thousand and one buildings all over the country but when the people get fed up with you and decide they want to get out, you've got nothing.

"I don't want to just bad-mouth the company . . . because there are some good things. I haven't hidden them. I keep telling them when I praise you and say you've done something well, it's praise deserved. However, because there are some good things, don't stop from trying to make things better. I keep them guessing. I threaten them, tell them to shape up or I'll write an editorial. I tell them I know people on the New York *Times*. They don't know. I know they'd like to get me out of their hair.

"I've been given offers. They don't come right out and say you can have this position, but they let you know that you could get up in the office if you want. As far as I'm concerned, this is all frosting on the cake. Fourteen and a half years ago I couldn't put my foot in the door, don't offer it to me now. I'm not going to be no token black. Even though I need the money, I won't sell my soul, and I'd have to. I'd have to conform and I don't know how.

"I tell them, go fly a kite. I don't want it. What would I be? A puppet that they could pull my string? What would my children think of me? Everything I do reflects back on them. Maybe when they grow up, I'll be scared for my job and do all those things. But right now I don't have to. I'm walking tall, ain't got a pot or a window but I'm walking tall. There's nothing they can offer me because I'm my own girl. You can't buy me.

"With everything I've said I know Western has done a lot for me. I've grown through Western. I had my ego thing at Western. I'm well known in the community. If I hadn't been at Western, it wouldn't have happened; I made my own little niche. I needed it at the time, I was a nobody because of the situation with my husband and it gave me a feeling of importance. When I leave, and I do intend to leave someday, I'll remember Western for what it has given me.

"The union has meant a lot to me too. I've gone to conventions and I'm well known by the heads of the International. When I go to the convention they all know me. They come down out of the stands and say hello to me and I feel good. They'll say there's Irene Lambert and it's nice. I've met representatives and senators.

"I think my husband resented all this. Here he was telling me I'm nothing and here I was going places, meeting important people who were telling me I was something. I thought I had gained self-confidence from this, but when John left me, I realized I didn't.

"I've had to grow up in the last three years. Now I know I have it. You know I got a lot of it through my children. They're behind me as much as I'm behind them. It's going to be very hard for me to let them go. I know that. I hope I'm big enough to let them go. I've had

them with me a long time, and you know how a mother is with her sons. I don't want to be forgotten. I fear this. That's why I feel very guilty about my parents, because I've forgotten them. I feel very close to my children. I think the breakup has made us even closer. When I'm upset they catch it. They know when I'm up and when I'm down.

"My youngest son can cook as well as I can. They all wash, cook, clean, iron, the whole bit. I get up tight when they don't do what I want them to. I don't want to come home and find this house dirty because I don't leave it dirty. I haven't cooked supper in weeks because I'm working overtime so supper's waiting when I come home.

"This is the type of house it is, probably a crazy house to other people but it's great to me. I don't think there's anything that I do that they shouldn't do, because they're men, and they're 100 per cent men.

"I'm very strict with them, partly because they could run all over me if I weren't tough with them. I demand top shelf from them. I don't like to ask them to do something more than once. They don't do it and I'm on them. I've hit them. I mean I've literally beaten them with a belt, even the biggest one, and he did not raise his hand to me. This is my way, it may not be right, but it's my way. I want them to obey my discipline, to do as I think is right. If they think I'm wrong we discuss it. If I think they're right I'll tell them, 'OK, I'm sorry I was wrong.' If they have personal problems, tell me, I tell them—talk to me, I'm the only one here.

"I try my best with them. I don't want to make the mistakes my parents made with me, but I probably make ten thousand others. I want them to have more than I had, but if they don't and they can have with their kids what I've had with them, they got it knocked.

"I'm not saying it's been easy. There have been a lot of problems. Even though there are more blacks in Haverhill than in Andover, there are still very few.

"My youngest, Yancy, he went through hell when we first moved in. He was the only black child in his school. He was NIGGER from the time he went in there to the time he came home, five days a week. Every day I'd be getting a note about how bad Yancy was. I didn't realize what he was going through until he finally came to me and told me how much he hurt. Oh, I felt terrible. I felt that I had failed him completely.

"Man, I went to that school and I performed. I gave that principal hell for being insensitive to the problems a black child would have in an all white school. I didn't leave until I knew he understood what they were doing to Yancy. Since then things have turned around. Yancy's gone from failing marks to being a B student. Once he knew that he could go to them when he was threatened and have something done about it, he was all right.

"I've had trouble with the police, too. Once they came into my house going after my son because some little white girl accused him of stealing her wallet. They didn't have a warrant or anything. Until they knew I wasn't going to take any crap from them, they were rude to me. I straightened things out and told them to get out of my house. As soon as they left I called up the chief of police and tore into him. I told him I got news for you, I don't know what you've been doing with other blacks, but your officers are not going to put their hands on my children and get away with it. I told him you ever want to come on my property you better have a warrant or keep your ass out.

"I'm a gentle peace-loving person, but don't mess with my kids. They know that if they're wrong, and I find out, you won't have to come after them because I'll bring them in myself. But when they haven't done anything don't touch them.

"Like I said, my children have been a joy to me. They have been a rock for me, with all the pain that John has put me through. I don't think my husband knows how bad he hurt me. He never abused me physically but he abused me terribly mentally. He had me to the point where I didn't think I was good for anything, and that's a terrible thing to do to another human being.

"I went to a black sensitivity program. I didn't come out in the group at first. I sat there and then finally everything came out. I was telling everything to people who could destroy me. Once it came out I could no more stop it than the man in the moon.

"I think out of the whole group I was the only one who got something out of it. There's a lot of bullshitting in those sensitivity groups but when I finished I felt as though I had taken a bath. I think that it's only since then that I started to heal, really.

"Now I want the headaches and the concerns gone. I want to help my children get started and then I want to find someone. I would like to love again, but I'm afraid to. I wish there were someone who was strong enough who could see what I need and give it to me, and it isn't things, it's time. No one has ever given me their time. My parents never gave it to me. They were too old. I never came first to anybody. My husband didn't give me the time I needed. My children, I do come first to them now, but I won't always come first to them.

"I want to come first with a man and I want to be his total. I don't know if there's any such thing, but I want it, really want it. When I see a couple that's really close, I'm very envious. Sometimes I wonder, when are things going to get better for me? When am I going to go to sleep and sleep a sound sleep?"

13. A SINGLE PARENT

CAROL LEAVITT—It took a couple of months before Carol and I started talking. My interest in interviewing her began one day when she told me, "I don't know how people get their impression of what I'm like, but I'm a lot different from what a lot of people in here think."

1969 was Western Electric's centennial year. It was the year that Carol Leavitt was selected in a plant-wide election as Miss WE Valley. Until the day of the election, the then twenty-nine-year-old divorced mother was unsure if she wanted the title. But standing on the stage, in the plant auditorium, waiting with nine other finalists, all doubts disappeared—she wanted to win.

What followed was a very busy year for Carol. In the year that she reigned as the Works' hostess and representative, she traveled as far north as Mount Washington, New Hampshire, and as far south as Boston. While in Boston, several state senators took her into the empty council chambers. She was allowed to stand at the podium and rap the gavel. She still has the souvenir gavel that she was presented.

All the occasions took up a lot of time. A day hardly went by when there wasn't an event to go to. It wasn't unusual for Carol to punch in at seven, primp up in the bathroom, be called off her job by 8:00, and not come back before 4:00 to punch-out.

Carol Leavitt got to meet many people, including Senator Edward Kennedy, that year. She went to many functions she would not otherwise have attended. Often she was in the company of the plant manager, Harry Snook, whom she remembers as a man who had class. He was the first big wheel with whom she had come in contact.

It was a good year, but Carol was ready for it to end. She felt the subtle pressure of needing to be constantly up on current affairs in order to give a good impression. She felt a tremendous responsibility to the company.

However, it wasn't easy going back to her normal schedule. She felt a confinement in her work routine that she hadn't felt previously. It was as if she had taken a step forward and now needed to step back. It took a lot of self-discipline to get back into the routine.

Carol Leavitt was one of the first people pointed out to me when I started work. She was always well dressed, as though she could go to a fancy restaurant after work. I didn't speak to her for a long time, but I noticed that men were always dropping by. Occasionally, I'd hear her talk about going to New York, about drinking at the St. Moritz or the Pierre. I thought of her as an attractive woman with expensive tastes.

Once I told her my impressions of her. She laughed and said, "You know, that must be the image I project, but I don't think that's me. I don't know how people get their ideas of what I'm like."

Outside the plant, away from the social evenings, Carol Leavitt is a single parent with two children, a daughter sixteen, and a son fourteen.

"It's taken me a long time to come to accept that I'm a single parent and can't be everything to my children. For a long time I felt 'Hey, I'm doing a real good job.' But then when my son started having disciplinary problems in school, I had to face the fact I'm a single parent, my son doesn't have a father to depend on. I can't be a father because I am not a man. So he had to face it and I have to face it.

"It got worse and worse. My son pushed me and pushed me and I was really frantic. One day, I couldn't take it anymore. I was screaming and hollering and I said, 'I'm sick of taking your bullshit, you either act right or I'm going to hit you,' and I did. I really hit him. After that it seemed like I was much calmer inside because I knew I couldn't do it all.

"I've been luckier than most single working parents, however. The kids have had a little more stability because we've lived in my mother's house. After my divorce I moved in with her, and I have not had the everyday money worries that I would have had if I were by myself. I don't have to concern myself with the running of the household. It's my mother's house and she takes care of it, she likes to.

"Also I've worried about the kids less, living in her house. I think the biggest worry of a working mother is where her children are, and what they're doing. My mother works the day shift, and I worked nights for many years. That way I could be with my children during the day, when they were young, and she was here at nights. My mother has taken a lot off my shoulders; she's the most unselfish person I think I've ever known. We disagree about how to bring up the kids, but she doesn't interfere.

"But, as much help as she is, being a single parent is still very lonely. You constantly realize there is no husband around to talk to when the children misbehave. And when the children do something and you're very proud, there's nobody to share it with. You're all alone all the time. When the children are in bed you're still all by yourself. Sure you may go out. I go out in the evening, maybe for a few drinks and dinner, but then I go home and I'm still all alone.

"I think for a long time I was too tough on myself. I was too aware of the image I was trying to present. I wanted to present a strong image, an image of someone who could always cope. Now I can see that was silly. I would try not to get mad at my kids. I can remember times when they would get me so mad. At times I thought I could kill them. At times I wanted to say, but I never did, 'Well, I didn't choose for you to be upset now, or for you to do this.' But no, I thought, they have precious little egos. They can tell you if they're upset but you can't tell them how upset they can make you, because you're not supposed to hurt their little egos. I had to get away from that. I had to realize I was hurting them by not showing them how I felt.

"For a long time, I was all the way to one side of the pendulum. I was the whole mother thing. Then I said the hell with this, I'm going to go out and have a good time, which I did. My children suffered and they started having problems. So I said I can't do this either, let me see if I can get in the middle, strike some kind of balance.

"That's really the big problem, trying to draw the line. How much should I take for my own needs? How much should I give to them? Where do I stand in my own life? One of the things I have to decide is how involved do I want to get with someone romantically. Right now I tend to stick with people who have a full life of their own, like I have. After I get home from work and take care of my kids I don't have a lot to give. I don't want any pressures from a man. I don't want anybody pushing me. I want to go out, have a good time, and relax.

"My kids tend not to get involved with my relationships with men. It's always been that way. Years ago, when they were young, and I was younger, I always kept my social life separate from my home. We lived in my mother's house and I never felt comfortable about having boy friends hanging around the house. I couldn't infringe upon my mother's hospitality, so I just didn't bring men around that much.

"Also my family, my brothers and sisters, and my mom and father, we've always been close. So on holidays, birthdays, anything, we've always been together. I've never really brought many outsiders into the family unit. So my kids were always very passive about it, they don't get involved with my boy friends.

"What I want from a relationship is still in the future. Right now my number-one concern is my children. As much as I love them, I will be glad when they grow up. Then they can make their own decisions, and I can make my own decisions. I can do what I want to do. That'll be great, a luxury.

"I used to think when they were little, My God, these kids are never going to grow up. I'm going to be tied up until I'm eighty-nine. And now I'm thinking my oldest, my daughter, is going to be sixteen and my son is fourteen. Soon they'll be making their own decisions.

"Like every other parent I had all kinds of dreams about what my kids would be. I've had to accept that I don't know how they'll turn out. I'd like my kids to grow up to be happy constructive people. That's all I'm going to ask. Before I wanted them to be something great. But now I realize they can't, their background is not conducive to them really being something.

"I grew up in a very small neighborhood. Everyone went to the same church, the same school. The neighborhood was the focal point of my life. You went to school, you went to work, and you got married, that was all. So at eighteen, just after I graduated from high school, I got married, and within a couple of years I had two kids and I was tied into a bad marriage.

"Now when I look back I think if I could only spare them the headaches and the pain. This is the hopeless part of being a parent, you cannot make your children avoid mistakes.

"I was hoping, this is one of my dreams, to introduce my children, especially my daughter, to a better class of people, to a nicer life. My Mary is very special and I'd like her to get the best out of life. I wanted her to realize that there is more to life than getting a man, getting married, having kids quickly and tying yourself down.

"It's taken me a long time to learn about other things. I like having nice clothes, meeting people with something to say, having dinners at fine restaurants and traveling. I am trying to interest my children in these things. For example, I just sent Mary on a ten-day trip to London. I had to strap myself financially to do this but I thought it might widen her view. When she came back she said, 'Well, I learned, I met other people. I saw the way other people lived.' I really don't feel she did. I think it all went over her head.

"I've tried to explain to her that the way you learn is to find somebody above you to learn from, you don't go below you, you go above you, but she rejects this philosophy entirely. She doesn't want any part of it. My children are probably going to stay at the level they are now because they haven't acquired those tools, the social skills to better themselves. I just hope they can find some happiness.

"My own immediate goal, my goal for the next few years is graduating. I'm definitely going to graduate from Northern Essex Community College. After that I don't know what I want to do. Maybe I'd like to break out, leave Western, do something else. I've never had the freedom to shop around and find out where I belong, what it is I wanted to do. When I came to Western twelve years ago I was twenty-three years old. I was divorced and I had two small kids. My mother and father worked for Western, and at one time so did my brother and sister. It was the simplest place for me to get a job. Really, necessity made me come here.

"I needed the security when I was younger, not just financially but personally. Even though I may exude confidence, basically I feel that I'm insecure. I was worse years ago. A little thing, a little rebuff from anybody, used to really hurt me. It bothered me when it shouldn't have. I used to let little picky things get to me. I used to want to be accepted by everybody. You can't. When I was younger I thought I could; I thought I could bend all sorts of ways. As I get older I find I can't bend as much as I used to. I can bend only in certain ways and I'm at the point where I am particular about the way I bend.

"For a long time I was insecure because I wasn't sure of what my

capabilities were. As I grow older I'm finding them. Unfortunately I'm now thirty-five years old and I'm still searching.

"I'm less anxious now about not knowing where I'm going. Maybe by the time I'm forty I'll be somewhere, I don't know. I used to think I was the only one who thought like that. I thought, What's the matter with me? Why can't I reach a plateau and stay there and be happy? Now I feel I'm much more sophisticated, more secure than I ever was. But I'm still searching, looking for fulfillment—what makes me happy.

"Maybe I'll get out of the factory. Do I want to stay at Western or do something else? Sometimes I think I'd like to try something else, but you have to have a financial base to really experiment. I'll tell you, I talk about leaving, I seriously consider it, but it would be hard. I probably make more working in this factory than the average woman does. This year I'll probably make around twelve thousand dollars. It would be hard to give that up.

"If I stay at Western I'd like to get ahead. For me that means getting into management. The woman's movement has helped open the way for women to get into management. This is really where the woman's movement has helped me.

"It really hasn't affected me at home. I'm too old to be really affected by it. I think my daughter is too old too. It has to start younger. I'm the type of woman, when I go out with a man, I want this man to make decisions. I want him opening the car door for me. I want to be led. I want to be told, 'Tonight we're going here,' and this or that. I'm still that way.

"I don't want to be equal at work, either. I want to be treated as a feminine woman. I don't want to wear overalls to work. I don't want to lift anything up. I don't care what they pay me, I don't want to do a man's job. But, let's say a job as a tester is open. If I can do it well, why shouldn't I have the job? If I study, if I pass all the tests, why shouldn't I have the job? I think as long as women are capable, if they have the background, they should have the chance.

"Now women are encouraged to do things, to do what they want. When I came in here it was different. The possibilities were there, but you were discouraged from doing certain jobs. Like in testing, they said, 'You girls could never do this job,' but we knew we could do it and we did. Lots of times we turned in a better performance than the men did. Eleven or twelve years ago I was among the first women testers doing technically oriented jobs. In fact, at one time, I was the youngest woman 36 grade tester at Western. Most other women weren't going to take time out of their lives to study electronics. I did, because I liked it. I felt I had an aptitude for it. I used to study on my own.

"As far as the grade system goes in here, I don't think there has been much sex discrimination. Being offered a higher grade job is based on seniority. Service is what counts. If you want to go into management, now that's a different ball game.

"You could get ahead if you were a woman, but it was awfully difficult. Now they're promoting women in here. The equal employment laws, and the company's concern for its public image, is opening up places in management. Here is where I think the woman's movement has helped to change things.

"This is the avenue open to me. The problem is, they don't operate on seniority on the management side. They have a complex system to determine if you're qualified. Qualified not only means college credits, you have to have the right temperament, an ability to make decisions, an ability to be a leader.

"You can't ever be certain that you'll be promoted. I think I have the qualifications. The only thing I needed was some college credits and I've been going to school for a couple of years now.

"I'll be honest with you. I hope I'll get promoted. My job is all right, and I think I perform it well, but I'd like to do better. I'd like to get ahead.

"I've been concerned that I haven't gotten a promotion yet. It hurts a little to see other people get moved up, while I stay where I am. I'm hoping that my chance will be coming soon. It's frustrating waiting, but there isn't anything I can do. I've gotten to the point where I try not to think about it. Instead I do my job. As I said, I think I'll get ahead, but maybe I won't. If I don't maybe I'll leave, who knows? Sometimes I wonder, do I want to spend my life working for Western Electric?"

SOFT
DRINKS
"WORK"
Beat Road Fatigue_
TAKE A "BREAK"

14. A SUMMER'S VIEW

Each year several hundred students apply to Western Electric for summer jobs. The company gives preference to sons and daughters of its own employees. In 1973 two hundred fifty students applied for summer work, and of the sixty-six who were hired, thirty-seven were the children of shop employees. They were chosen on the basis of the length of their parent's service with the company. Three of these summer employees reported to work in our shop on June 5. Each had a father who worked as a layout operator and had more than seventeen years of company service.

People seemed to treat these three summer workers, who spent most of their free time with one another, with special tolerance. Hard work was appreciated, but poor work or a lackadaisical attitude seemed to be accepted. As one woman said, "Look, they're only kids, they're only here for the summer." Another woman seemed to express a commonly held opinion when she said, "If their job here this summer doesn't teach them anything else, it should teach them how important school is." Shaking her head, she added, "I only wish my Billy would learn that."

A rather maternal attitude was taken toward the three students by the women in our group. Nevertheless there were still differences in how the three were perceived. Neal Webster was seen by practically everyone as the most responsible.[1] He was, as Mary Smith remarked, "a diligent worker." As she said to me, "He does his work. He's a good boy."

People did not say much about Alice Janecki, except that she was shy. I rarely talked to her. In fact the only vivid memory I have of her comes from something she did on her last day of work. She came downstairs after lunch and told a group of us that she had an urge to push Larry Tuttle's feet off his desk (he always took a nap at the end of lunch). Neal, Joey, and several others of us dared her to do it. She looked at Larry, giggled, and with her right hand brushed his feet off their perch. Larry woke up, looked around, and saw Alice halfway up the aisle. He gave her a look of disgust, closed his eyes again, and put his feet back on the desk.

[1] All the names in this chapter have been changed.

Joey Tamosi seemed to irritate a number of the workers. Fay Casey told me more than once, "He thinks he's so damn smart, thinks he knows everything. He doesn't, he's just a spoiled kid." John Keker said, "I don't know what it is but the kid just rubs me the wrong way. Someday he's going to open that big mouth of his and someone's going to put their fist in it." Arlene Dowd took a much more philosophical view of Joey's behavior, one that was very close to my own feelings. I told her one day that I liked Joey but never understood why he seemingly went out of his way to piss people off. She said, "He's just young. He doesn't mean half of the things he says. Basically he's a good kid."

Alice Janecki, Joey Tamosi, and Neal Webster are very different people. I tried to find out from them what the summer had meant to them and what they had learned.

Alice Janecki's hands fidget nervously as she talks of saving enough money this summer to buy a pair of Helix competition skis, to share in a ski lodge with twenty other kids from college, to ski Loon Mountain, Jay Peak, Wild Cat, and Killington. She is even thinking of going skiing with her boy friend in Colorado. She has already begun preparing for the winter by getting up at 4:00 every morning in order to run a mile or so to strengthen her legs. The more Alice talks about skiing, the more relaxed she becomes.

"I love trick skiing," she says. "I'm really getting to be good. I only got interested in it a couple of years ago and already I can do all sorts of tricks. I can't wait to go back up to the lodge. Last year was my first time doing something like that and it was terrific. At home I can talk to my parents, but it isn't like when you talk to kids. I don't know how to explain it, but I've really opened up with these kids. Even our chaperons are two young teachers from school, and they sit around and talk to us. We stay up real late, after skiing sometimes till three or four in the morning, talking. I just love it."

Alice had another job at the beginning of the summer; a girl friend of hers who was working for a plastics firm got it for her.

"I hated the place," Alice says. "I couldn't believe how filthy it was. It wasn't hard work, and I didn't mind the pay [$1.75 per hour] but I couldn't get used to the filth. The job was frustrating. It was horrible.

"So one day, I think it was on a Thursday, less than a week after I began work, I told my girl friend who was working right next to me that I was leaving. She looked at me but she didn't say anything. I said, 'Just tell them I've quit and they can mail me my money.' It wasn't anything I had planned. I was frustrated and decided to leave. I put down my tools and walked out.

"I was lucky, though, the next week I found out about the job here at Western Electric. When I started at Western, I was also working at Salisbury Park [a local beach amusement area]. I kept both jobs for a couple of weeks but it got to be too much for me so I quit the job at Salisbury Park.

"The work here at Western Electric hasn't been bad. The first week I was scared. Everything looked so complicated I was sure I'd do everything wrong. But after a week, I got to know the job pretty well and it got so I could do the boards pretty well.

"As the weeks went by, I got more used to things. I got bored some of the time. I think I hated working in here the most on a nice day. I almost couldn't stand being in here when it was so nice out. I'd just sit there thinking about hiking or skiing. Sometimes I'd think of doing something crazy like I'd look up at the little fan over my desk and I'd think of putting my finger in it. I never did it, but I had the urge to do it a lot.

"I found the days would go by pretty quickly if you could talk and work. When there was no one to talk to, I'd make up all these little games. Sometimes I'd try, as I was cutting the leads on the back of my boards, to hit Joey or Neal with them. [She laughs] I actually got to be a pretty good shot. Sooner or later, though, they'd tell me to cut it out and I'd have to think of something else to do."

"Did you work during the vacation?" I asked.

"Yes. When Al asked me if I wanted to work, I said I would because I could use the money. I think I liked working during those two weeks of vacation better than anything else. We worked in a small group, and everyone was real close, including the supervisor and the engineers. There were maybe twenty-five of us. We all took lunch together, even the supervisor. We'd go across the street and get a pizza or some subs. People really sat and talked to you. Our supervisor let you know what he expected and we tried to do it. Everyone was really nice, and you didn't feel you were being watched every minute. If the supervisor took a break, well, I felt I could take a break. I really enjoyed working those two weeks."

"Have you learned a lot from the people you've worked with over the summer?" I asked.

"Well, most of the time I'm with the other kids in here, because I feel I can be freer with them. Some of the older people warned me that if you tell people anything in here it will be spread all over and that I should keep my mouth shut. That made me sort of shy, but a lot of the people have talked to me, especially about school.

"A lot of the women tell me how lucky I am to go to school. They tell me they'd love to be going to school. I guess they're right.

"This summer made me realize how much I want to go back to school. I think I'll be more serious in school and not fool around so much. I can see now how much having a college education means.

When I was a senior in high school, I didn't really think about going to college. I just thought I'd graduate from high school and get a job.

"My parents had a talk with me one night. My father told me that if I went to a communiy college he would pay my tuition for two years while I got my associate's degree. He said if I wanted to continue my education after that I had to pay for it myself, but he wanted me to go to school if I could.

"He asked me what I thought about what he'd said. I was really surprised; I guess I never expected something like that. When my dad asked me again what I thought, I told him it was fabulous. I got some applications and I got into NECO [Northern Essex Community College].

"I guess I haven't worked as hard as I could there, but now I'm really going to work when I go back. Now I want to get an associate's degree in business, and maybe go on in school and get a degree in child care.

"This may sound corny but working here, I've learned to respect my father. I mean, I love my father, but I never imagined what his work must be like. Working here has shown me what he must have put up with over these years. I know it has changed my behavior at home. Sometimes my dad used to come home and ask me to do something. I might do it, but sometimes I felt a little resentful. I felt I was tired too, and why should I do it. Now I'd do what he asks. It's not that I pity him, but I realize what he's done so that I could have a good life, and I guess I appreciate that more now."

"Would you work here again?" I asked Alice.

"Yeah," she said, "I think so, if I couldn't get something better. If it was a choice between working in a restaurant and working here, I guess I'd work here. I mean where else could you make this kind of money for doing this kind of work?"

Joey Tamosi strides rather than walks, carrying his five-foot-eleven, one-hundred-sixty-pound frame well. He often comes to work in cutaway tee shirts, which expose his well-developed arms. The gait and the physique are cocky, arrogantly sure, and yet there is the odd Dutch-boy haircut. When Joey talks, one is immediately aware that a somewhat insecure boy stands inside this man's body. His speech is full of long words, which more frequently than not are incorrectly used. He awkwardly grabs for the spotlight in a group conversation. He defends himself with the putdown. Paradoxically, he seems both acutely aware of the often disastrous effect of his approach and shamelessly unaware of it.

Alone, away from the crowd, Joey talks about himself and his feelings without using either big words or putdowns. I ask Joey what he thinks of the work he has been doing here.

"The job is nothing," he says. "It's incredibly boring. There's no challenge to the work. Almost as soon as I came in I knew I didn't like it. I sized up the situation and decided I wasn't going to kill myself. I'd do enough to get by and that's it."

"Do you think that Jack [our supervisor] is aware of your attitude?"

"Yeah, I know he is. He hasn't ever said anything, but like he put both Neal and Alice on the big boards. He didn't put me on them because I guess he knows I didn't give a damn, and he can get more work out of them than me. It isn't that they like the work any more than I do, they're just more submissive.

"I was like that the summer before this when I worked in a clothing factory. I was younger then, more submissive. I was scared and aware of the bosses all the time. I worked, not because I liked it, but because if someone gave me an order I did it. Some of the kids would sneak off to the back and catch some sleep, but not me, I never did that. But as I matured I knew I didn't have to put up with that crap anymore. See, I went to Catholic schools, and you learned about following orders, and that's how I acted, but now I know better."

"Was the place any different than you expected?" I asked.

"No, not really," Joey says. "This past year at Merrimack [Merrimack College] I had a second division history course. Our teacher taught us about the dehumanizing aspects of industrialization. We learned about punching a clock, the monotony of work. In the class the teacher pointed out that in the old days it was better to be a farmer than a worker, even if you made less, because you'd be happier.

"This place is a little cleaner than I expected, I'll grant you that, but it's monotonous. Basically, the work doesn't change, day after day. I found I had to play all these mind games to keep going. Sometimes my mind will go off—on a tangent. I'll think that I'm sitting on some rocks that are tarnished, or maybe I'll think of the glitter of a lake. I'll think I'm all kinds of places but here. Other times, I'll talk to someone near me, I'll try to figure them out.

"Sometimes I do stuff just to get a reaction. Like the other day, after I emptied the wastebaskets in the back row, I left them in the front row. Well, a couple of the women got real mad, wanted to know where their baskets were. People were yelling at me, it was vicious. I really didn't think they'd get that mad over some dumb wastebaskets.

"I can't figure out some of these people. This one woman, Fay, won't even talk to me. And the other day this woman came up to me and she asked me, 'Well, Joey, have you done anything today?' Well, I was just kidding around with her, and I waved her off, you know like this with a little wave of my hand. Well, she turned around and she looked at me with real hatred and then she told me she didn't want me to talk to her ever again. So I guess some of the women don't like how I act. But I can't really understand why. I'm not asking them to act like me. I'm trying to be honest about the way I feel. I don't try to hurt anyone.

"Sometimes I think I talk too much and that I kill myself with my mouth. Maybe that's something I've learned this summer. That I ought to have kept my mouth shut. Maybe I should say things in a different way. Somehow I don't get my meaning across sometimes. I know some of the time it's because I'm speaking over people's heads, but other times it's because people have already formed an opinion about me and they don't give me a chance. There's really nothing I can do about that, is there?

"I'll give you an example of what happens. The other day we were sitting outside and a woman from another department came by and said to me, 'Joey, what else do you do but look handsome?' Well, Jean said, 'He doesn't even do that.' See, she was putting me down. But I let it go because I think she was putting me down because she likes me.

"I can tell from your look, Dick, that you think I'm wrong. But with all the girls I've known, I've been wrong maybe once about saying one liked me. Come to think about it, I can't even think of one time I was wrong."

"What is it that you want to do, then?" I ask.

"I'd like to go into special education. You know, I've worked at an orphanage for the last year, and I love it, really I do. Those kids are really honest. Let me tell you about something that happened there the other day, because it sticks in my mind. These two kids, Gino and John, were standing around. Well, Gino ripped this bat away from John and just went off with it. I went up to Gino and said, 'Why'd you take Johnny's bat?' He looked at me and said, 'Because I wanted it.' I thought, Wow, that's really honest. He didn't give me any phony answers, he just said because he wanted it. I mean, that was really honest. I asked him, 'What if I want the bat and take it from you, that isn't right, is it? Why don't you give it back to Johnny?' He said OK and gave it back. Well, that honesty meant a lot to me, and I'd like to work with those kids, or kids in school who are in special classes and don't have enough people who care about them. I want to do something important, something to help other people.

"This summer has been good, because it's made me think about myself and how I've changed. I used to be pretty submissive. But about two years ago I grew about four inches and put on about twenty pounds. It made a lot of difference. I could tell from the way my peer group treated me. My size helped give me a lot of confidence."

Just as height has been important for Joey Tamosi, it has been important for Neal Webster. Neal says he's five foot six, but he may be an inch or so smaller. He is acutely aware of his size.

"You know," he says, "I used to pray when I was younger that I would be taller. I was always smaller than my friends. It's made me shy. It's especially affected me with girls. I can't really impress them with my size, so I have to get to know a girl pretty well before I'll ask her out. I guess I want them to know I have a good personality, to overcome my physical impression.

"Being short really has its disadvantages. Now they have some laws about discriminating against people because of their height, but people are mean to you and they don't even mean anything by it. Even my younger brother. For crying out loud, he's only sixteen and he's almost six feet tall. Sometimes he'll say something like, 'How's my little brother doing?' I never say anything but it gets to me. The only thing is he's a little clumsy, so when he says stuff like that I suggest we go out and play some basketball, one on one. I always beat him, and then I say something like, 'I'd rather be small and quick than big and clumsy.'

"Anyway, I think I deal with my height better now than I used to. I mean, for crying out loud, I'm not going to get any taller. I've tried hard to do things well to make up for not being tall. I found out that I could do pretty well in school, and I've worked hard at it."

"Someone told me," I say, "that you were first in your class in high school."

"No, I don't know who told you that but I was third in a class of two hundred twenty-three. I graduated with a 96.5 per cent average."

"How did you select Merrimack College?" I ask.

"Actually I had thought of applying to MIT or Harvard, or places like that. I talked to my parents about it, because they couldn't afford to send me to college. I've got two older sisters, and one would have liked to have gone to college, but neither did. We couldn't afford it and my mother felt it wasn't so important for a girl.

"Anyway, money was a big thing. My father has had to hold two jobs to have us get by. He's worked very hard. But I told them that my advisers thought maybe with my record I could get a scholarship to one of these colleges. Well, my mother said she didn't want me to go away from home. Kent State had just happened in my junior year of high school, and she said she would always worry about me if I lived away from home. She said she'd be scared all the time, and my being away might cause her to have a heart attack.

"She asked me couldn't I find a place around here to go to school, so I picked Merrimack. It has a good physics department and I've got two scholarships for $2,800. That means I actually have a little extra after paying the $1,900 for tuition and $150 for fees.

"Occasionally I feel gypped, but I plan to go away to graduate school to study astronomy. My mother says she doesn't worry so much about me now that the campuses have quieted down.

"My advisers think that I should be able to win a scholarship; I hope I do, but I'm trying to put away enough so that I could afford a year away at school even if I didn't get a scholarship. What with my scholarship, summer work, and work during the year, I hope to have $4,000 saved up by the time I graduate.

"I had my first summer job as a janitor when I was thirteen. I've had other jobs during the summers; once I worked for Kentucky Fried Chicken. The last couple of years, starting while I was a junior in high school, I worked for DeMoulas [a large local grocery store]. I began as a sacker. I worked the whole summer and then during the school year I worked on Friday and Saturday night. They started me at $1.80 an hour and I just quit this past spring after three years and I was making only $2.10 an hour.

"This past year they made me assistant cashier. Well, I liked that, but I thought they should be giving me some more money if they wanted me to take on the extra responsibility. I asked them about a raise several times. They gave me a runaround. They didn't give me more money, but they kept telling me to cut my sideburns. They weren't real long, but they wanted me to cut them back to my ears. I figured they had no right to ask me that. I never did cut them. Finally I quit.

"I needed a job for the summer. I put in applications at Gillette, Raytheon, and Western Electric. Raytheon said no right out. Western Electric and Gillette put me on the waiting list. I heard from Western Electric first."

"Would you work here again?" I ask.

"Now that the summer's over, I can say that unless I found a job in my profession, you know, something to do with astronomy, I'd take a job here again next summer. It's not that I found the job all that exciting or interesting. If anything, it's boring. But the pay is decent and the people were OK.

"I didn't find it that hard to get through the summer. Once in a while I'd start daydreaming or wishing I was somewhere else. But I realized I was here to make money. If I didn't have this job, what other job would I have? There wasn't really anything. And if I wasn't working I'd be bored, and spending my money, which I really can't afford to do.

"So I figured I'd put in my eight hours. I'd be out of here each day at three. A lot of the time I played this game with myself. Each day I'd try to break my previous day's record for work. It made the time go quicker. I found the busier I was, the faster the time went. When I worked, I tried to work hard. That way I didn't feel bad when I got up to go to the bathroom or for a walk. I think I handled myself maturely. I had to look at the situation, and had to decide what to give. I was really on my own. I figured there was an obligation which was important. I mean, for crying out loud, you've got to respect their rights. You can't take their money for nothing. I mean, they aren't paying you to sit around. I figured I kept my part of the bargain. They gave me a decent job at a decent wage.

"But I'm glad the summer is over. I learned this summer how much I really like school. I'm anxious to get back. This is the first time I've had this kind of regimentation. I mean, the alarm goes off at five-fifteen. At six-thirty a bell goes off, and you're supposed to begin work, and there are bells all day long. Really, I wouldn't mind staying in school for the rest of my life."

"Did you like working in this department?"

"Yes," he says, "even though I was in here for a pretty short time I did work in a couple of different areas, and I'm glad that I worked in our area for most of the summer, because it was a miscellaneous area, there wasn't so much talk about rates, and work was less repetitious in this area.

"It wasn't that way in one of the areas I worked in. You know, Jack lent me, Joey, and Alice out to help Dennis Murphy, who had a big shipment and needed some extra people. I didn't like it there from the first day.

"I was sitting next to this woman doing some work. She talked a lot, because she told me there hadn't been anyone in her row for a long time. We talked but we both did our work. Anyway, the next day the supervisor called me over and told me he hadn't brought me over here to bullshit and slow down his people. Well, I could have said it wasn't me who was doing the talking, but he scared me a little and I didn't say anything.

"I worked hard for the next couple of days. The supervisor called us over again and told us that he had production to get out and we weren't making the rate. I'll say this for Jack, as soon as he heard what was happening he pulled us off the job and took us back.

"I didn't like working for that man. He pushed you all the time. He was always talking about the rates. I'm willing to work hard, but not like he wanted. I know if I had worked there all summer I wouldn't want to come back.

"I also had the chance to work for another supervisor. Since we weren't entitled to any vacation I decided that I would work for one of the two weeks that the plant was closed. We were sent over to help Jerry Malone right on the other side of our group. I liked him. He explained on Monday what we were to do, and said if we could complete sixty-four boards by the end of the week he would be happy. You know, he didn't say what the rate was or anything, just what he'd like us to complete. Well, I got my sixty-four boards finished early Friday, and I was pleased with myself.

"He told me he was real pleased, and that he'd always have work for me if I needed it. Originally I had planned to take off that entire second week, but I worked for him another three days. The place was a lot more relaxed during vacation."

"How about our area?" I ask again.

"Like I said, I liked working in this area. Jack never told us we had to make rates. You could tell from the way he acted that he knew how you were doing, but he didn't push you. He left it up to you. Even today, my last day. He called me over to talk to me. He told me he couldn't really judge my work in this period, that I'd have to be the judge. He told me that I would know whether I had done a good job or not. I think that's right."

"What did you learn about the work and the people this summer?"

"It's really the first time I've worked with grownups, and especially grownups who treated me like a grownup and not like a child. I mean, the only really older person I knew before was my grandmother. The women who worked at DeMoulas always made you feel like a kid and treated you that way. Here, they seemed to treat me just as another person. They'd ask me about courses and stuff. It may sound silly, but it was like a bunch of guys hanging out. They would sit and talk and you could join in the conversation.

"I learned that most of these people want to do a good job. For me it's a summer job, but for these people it's their lives. They want to do well, I know they do. You could tell how disappointed they were when they only earned a 5 per cent bonus last month. Even though they were paid more I know from what a lot of them said they were really upset.

"It reminded me of something my dad told me. Once a long time ago when they were starting a new job, they asked him to make the rate, which was twenty boards a day. He tried his hardest and just did make twenty. The next day they asked him to do twenty-one boards. He and all the other people never did make even twenty boards again. The best they did was eighteen or nineteen boards. He said no matter what they asked, if you did it, they'd ask for more. So, instead, just do what you think is fair, because a lot of them don't appreciate what you do."

"Have you talked to your dad much about his work here?"

"My dad is kind of quiet. We don't talk all that much. Even in here he doesn't talk much. I've gone over to visit him in his group, and people will be talking during a break, and he'll be off reading. Sometimes we don't even talk on the way in. He used to tell me how he wished he'd had the chance to go to college. His family was poor and he had to quit school at sixteen. Even when I was a kid he often held two jobs. We never had much, but we did OK, he saw to that. I don't think he loves his job, but I know he doesn't dislike it. I think he thinks he's doing pretty well. For him, anyway, it's a good job, maybe the best he could get. Anyway, who am I to judge him?

"I tell you one thing, I appreciate my dad a lot more after working here. I've always tried to help out at home. I've always given my folks some of what I've made. This summer I gave my mother thirty dollars from my weekly paycheck. When I was working during the year at DeMoulas I was clearing thirty dollars a week and I gave her eight dollars. It made me feel like I was contributing. My sister, who isn't married and lives at home, still gives my mom something every week.

"I came to like a lot of the people. Some of them are really bright. I learned a lot from them, but one thing I could never figure out was their attitude toward work. One way or another, many of them would tell you they didn't like their job. I sort of believed them, but when I watch them they seem not to mind it. They seem to have their friends and do a lot of talking. It doesn't seem to be really drudgery. I don't know, if I hadn't talked with them I'd say some of them seem to like what they're doing.

"Anyway, I'm glad to be going back to school. You know how a lot of people say everyone should spend time in the service, well, I think everyone should work in a factory like this for a while."

15. SOCIAL EVENTS

Company-connected going-away parties, retirement dinners, sports banquets, Christmas parties, dances, picnics, and trips are all social events in which large numbers of shop employees participate. These social occasions are infrequently mentioned but are important ways in which workers share in each other's lives. Social activity binds workers to one another and makes the company a more human less remote and alientating organization.

A change of shift, movement from one shop to another within the plant, or leaving the company, all call for a going-away party. Unlike many office parties that consume the better part of an afternoon, or fill up an out-of-office lunch hour, these parties are held in the work area and squeezed into the ten-minute morning break.

The going-away party is a highly ritualized event. Everyone in the group expects a collection to be taken, a card to be signed, and a party, including gifts and a cake, to be held. The company doesn't sanction these parties or the collections taken for them. However, it tolerates the collections and allows the parties to be held as long as they are limited to the morning break.

Within this framework the group determines the nature of the party. These parties are common to all work groups but vary greatly in terms of emotional content, depending on the cohesiveness of the work group and the feelings about the worker who is leaving. While I worked in the plant I attended four of these parties and saw several others.

Nora Hartwig,[1] who left the company when she was seven months pregnant, was given a party by our group. About a week before Nora left, Irene, a girl who worked behind me, came over and said, "Here," giving me an envelope. "It's for Nora, give what you want, most people are giving a dollar." Irene didn't wait for me to get my money. Instead she told me to pass the envelope on and not to let Nora see it. I put my dollar in and passed the envelope up to Dottie, repeating what Irene had told me. I watched the envelope as it circulated in the rows in front of me. At each place the previous person would leave to allow the person to give as he or she saw fit.

[1] The names in this chapter have been changed.

That was the last I heard of the party until two days before it took place, when I overheard Irene describing to Laurie, in great detail, all the baby clothes and the cake decorated with a small stork and a pair of baby shoes that she and Arnold had bought for Nora with the thirty dollars that had been collected.

On the day of the party, a Friday (Friday is usually the last day a person is with a group and therefore the day on which the party is held), the festivities began when one of the women pinned a corsage on Nora. Loving the attention but slightly embarrassed by it, Nora went back to work with her cheeks flushed red.

At 8:50 both Irene and Laurie disappeared and brought out paper plates and forks, as well as a huge cake. By 8:55, five minutes before the nine-o'clock bell was to ring, all the women in the front area had surrounded Nora and the cake.

The nine-o'clock bell soon rang and a few people, mostly men, went to the back area for their morning smoke, missing the cutting of the cake, but unspokenly assured that pieces of cake would be saved for them. Nora made the first cut into the cake. Then another woman took over as she turned her attention to the pile of gifts.

At 9:10 the buzzer went off again and people returned to their work. In that ten minutes a card had been read, a cake had been cut, and gifts had been opened. Before the day was over, several of Nora's friends from other parts of the plant had come to say good-by, to admire her presents, or to bring ones of their own.

At 2:55, while tools were put away, people began kissing Nora good-by. Nora had smiled and laughed throughout the day. Standing there being kissed, she suddenly began to cry.

The cake, card, and present are common ingredients to almost all going-away parties. There were no presents at only one party I attended, the one for the three summer students. To me that was a way in which the group symbolically, though perhaps unconsciously, showed that these summer students were not considered a real part of the group. That's why it was important to me that at my going-away party, along with everything else, there was a present. I was given a slide viewer, something I needed and didn't have. More important to me than the gift was the card. Nearly everyone in the group signed it with some sort of personal remark. The remark that touched me the most was one made by Vicky Wrigley. She signed the card, "I'm glad when you passed by this corner you decided to stop," a take-off on the title of a previous book of mine, *Next Door, Down the Road, Around the Corner.*

The going-away party is the rare social occasion that takes place in the plant. Most do not. The company picnic is an annual social event with a long history. The Merrimack Valley Works quickly outgrew the company picnic. Instead of a company picnic there is now an outing day at nearby Canobie Lake Amusement Park. For one dollar you not only get into the park, but entrance to all of the more than twenty major rides.

It struck me as a great opportunity for people in my area to get together with each other's families. Workers are typically introduced to each other's families through stories and photographs. Pictures of husbands, wives, and children, as well as of weddings, vacations, and births, are in almost constant circulation. Over time one inevitably begins to share in the tragedies as well as the joys of fellow workers' families.

When the outing was announced I waited a few days, hoping that several of the people whom I liked would say that they were planning to bring their families. When no one said anything, I suggested that a group of us might take ours. To my surprise, there was basically no interest.

This reaction made me realize that to a great extent there is only a ritualized interest in each other's outside lives. Friendships are made, but many relationships don't go further than sharing time at work. I bought five tickets and planned a picnic day with my wife, her sister, and another couple.

Saturday, the day of the outing, was cold and raw. Rain threatened most of the day, yet when we arrived at noon the park was already jammed. There were long lines of children in front of each ride.

We were unsuccessful in our efforts to find a warm place to eat our picnic lunch, so we ate quickly on one of the park's picnic benches, huddled together against the cold wind. After lunch we spent most of the afternoon roaming around the park, stopping to take this and that ride and to eat candy apples and cotton candy.

I don't enjoy most amusement park rides. Since the rides were free, it made them deceptively attractive. After saying no several times I finally agreed to go on the roller coaster. I couldn't help remembering all the times as a child that my older brother had made me go on the roller coaster at Coney Island and how much I had hated it.

My friend Ramsey and I settled into one of the cars. As we were comfortably locked into the seat, the car slowly began its way up to the top of the first drop. I knew already that I had made a terrible mistake. The first drop began, Ramsey's mouth opened with a laugh, and my mouth closed, my Adam's apple bobbed, and I prayed for the ride to end.

I contented myself the rest of the afternoon with encouraging others to take rides as I watched.

As we left the park I realized that I didn't feel connected in any way with the thousands of people all around me. Once during the afternoon I saw someone I knew from work with her famly. We stopped and talked for a few minutes. Aside from that I felt no connection with the group I had shared the day with. It was like being at an amusement park on a normal day, although the park had been closed to everyone except ticket-holding Western Electric employees, their families and friends.

That doesn't mean that people didn't have a good time; most seemed to be having a very good time. For the kids it was a day of amusement-ride gorging, a day on which a child didn't have to worry about the price of a ride. Nevertheless there was little mixing. The vast majority of people were isolated in family groupings. Most Merrimack Valley shopworkers relate to the company through the small work group. At Canobie Lake there was no small definable group to relate to. Being a Western Electric employee wasn't enough to make it a meaningful shared experience.

Western Electric has outgrown the company picnic. What remains is a sanctioned, ritualized event called an annual outing. For me, and

for the people I talked to, it had less meaning than any other social event I've attended through my connection with the company.

The Great Boat Ride

Every so often notices are posted on the bulletin boards throughout the plant and forms sent around announcing all sorts of trips, with special prices for Western Electric employees and their families. The trips range from a week or ten days in Acapulco, a Nassau cruise, a visit to Florida's Walt Disney World; or Las Vegas, and weekend trips to New York City, Toronto, or Nova Scotia.

I had heard several people at work talk about these trips and how much they had enjoyed them, so Eileen and I signed up for a weekend trip to Nova Scotia along with two other couples from my work area. For eighty-nine dollars per person we were to get bus transportation to and from Portland, Maine, passage from Portland to Nova Scotia and back on the *Prince of Fundy*, a room going one way, a smorgasbord dinner, a hotel room in Yarmouth, lunch and dinner on Saturday, and free entrance to the hotel's lounge and show.

We sent in our deposits and waited for our tickets. A week before the trip we received a packet with tickets for our baggage, a colorful brochure about the beauty of Nova Scotia, and a detailed itinerary about where we would stay and what we would do. It all sounded very exciting.

Several women at work, who had been on similar weekend trips, asked me if Eileen was getting a new outfit. They talked about the clothes they had gotten for their trips and the adventure of going to a new place with other people.

By the time we met our friends at the company's parking lot, we were filled with anticipation. Like us, everyone appeared to be with a small group of friends. The buses, which were scheduled to arrive at 6:00 P.M., didn't arrive until 6:45. The long wait in the cold, to my surprise, didn't seem to dampen the general good mood.

When the buses finally arrived in Portland, we discovered that the *Prince of Fundy* was late. Again we had to wait, this time for more than an hour before we could board the ship. Waiting, we were to discover, would be a common thread throughout the weekend. Our group of 125 was quickly absorbed by the 500 or so other people waiting for the boat, which didn't pull away from the dock until close to 10:30, nearly seven and a half hours since I had put my working tools away for the day.

We weren't underway five minutes before the first crisis developed. We had been told that we had a free dinner coming. It wasn't clear whether that free dinner was to be this first night or Sunday. Several of the men scurried around looking for the tour guides, but they weren't much help. This was their first tour. I was beginning to become annoyed with their incompetence, but most of the other people, who had been on other trips with this tour company, took it in stride. I heard things like, "Well, they're only young girls," or, "We've traveled with this company and it's a good outfit."

Anyway, we found that we couldn't get the smorgasboard that night, so we abandoned eating in the expensive restaurant and settled for the cafeteria. It was hard for the six of us from my work area to eat together because, unlike us, many of the other passengers didn't have rooms and therefore had "staked out" the cafeteria benches for sleeping.

Saturday began early with a knock on the door from our friends, a giggle, and the remark, "That's enough of that, it's time to get up for breakfast." As it turned out, the boat was delayed and we wouldn't be getting into Yarmouth until 10:30 A.M., so there was plenty of time after breakfast for picture taking and relaxing on deck. The boat finally docked. After clearing customs we were taken to our hotel in an old school bus.

I'm not quite sure what I had expected Yarmouth to be like, but whatever I had anticipated, it was disappointing. The shoreline and immediate surroundings were more bleak than charming or rustic. Our hotel was not some grand old but declining building, as I had hoped, but a modest, motel-like five-story brick building.

Although the hotel was expecting us, the room assignments had been mixed up and we had to go through a complicated procedure to be reassigned rooms. Eileen and I left our suitcases in our room, where we had a view of the air vent from the adjoining building, and went out for a walk. We came back in time to join our friends for lunch. We soon discovered that the tour scheduled to begin at 2:00 had been pushed back to 3:15. The tour finally began at 3:30. The group split up into three buses. Our guide was a young eighteen-year-old college student whose father owned the bus we were riding on. From the beginning the tour was a disaster. The boy repeatedly pointed out such interesting things as: a boat sitting on the side of the road, a house that had just sold for $25,000, the bowling alley where the best grass in

Author and wife

town was sold, and the road to the site where they had been going to build Yarmouth before they had changed their minds.

Strangely enough, no one seemed upset. In fact, the tour and the guide became a big joke. Whatever the boy said sent everyone into fits of laughter. We began to feel like a group. We made one stop at a lighthouse and everyone hurried off the bus, took their cameras out, and started snapping away, here in front of the rock, then in front of the lighthouse. "Would you mind taking a picture of me and my wife?" "No, sure." "Thanks, want us to take a picture of you?"

Our big stop of the day was to be a summer resort, supposedly a big tourist attraction about thirty minutes outside of Yarmouth. On the way out of town the bus passed a factory, and the boy said, "That's our biggest factory." He continued, a note of pride in his voice, "It has over five hundred workers, and two shifts." That remark filled the bus with laughter as we whispered to each other about how small the place was compared to Western Electric. Finally one man said, "You think that's big, we work in a place where there are ten thousand workers and three shifts." People kept talking about that remark until we arrived at the summer resort.

Our guide kept up his patter, pointing out the house where the owner lived, telling us that the resort, which looked deserted, was quiet. Finally the bus stopped at the resort's general store, a most unimpressive, deteriorating wooden structure. The bus waited for forty-five minutes as people picked out an assortment of knickknacks, memorabilia of our brief Nova Scotia trip.

Secure that presents for children or friends, hadn't been forgotten, everyone settled into the bus for the ride back to the hotel. In spite of the absence of anything of real interest on this so-called tour, everyone seemed to be having a grand time. People began to point out things on the road, such as a barn, or a wheel, or a shopping center, which sent others into more laughter.

Several times someone pointed to a store and said something like, "You know, they have five employees, and two shifts." We shared a feeling of superiority in working for a big company.

That night everyone was dressed to the teeth. All the new outfits were worn to dinner in the hotel's dining room. The glow of the afternoon carried into the evening and people said hello to those they had shared the afternoon bus ride with, and settled down to dinner with their friends. We all discussed how boring the tour had been and yet how much fun we had. After dinner most people took advantage of their tickets for a free entry to the hotel's lounge.

We had been told to be ready to leave the next morning at 8:00 A.M. Needless to say, we didn't actually leave until 9:30. On a windy morning, our bus drove us right onto the boat. When the boat first pulled out of Yarmouth harbor, many people tried to stand in the stern as they had on the way in, but it was even colder and windier than it had been the day before. People slowly began to move inside.

The trip home was filled with wandering around the boat and enjoying a leisurely smorgasbord dinner. Ten hours after leaving the Yarmouth docks we were back on the buses heading to North Andover. I realized during the bus ride home that we had seen almost nothing of Nova Scotia, or Yarmouth for that matter, and yet it had been one of the nicest trips I had ever taken.

The important element had been the trip, and not Nova Scotia. For many people at work, a company-connected trip is a long-awaited adventure. These trips include leaving responsibilities for a while, getting new clothes, carrying cameras, buying gifts, and sharing an adventure with other people.

At the beginning of the Nova Scotia weekend people had been isolated. That is, although most people were with friends, there really had been little interaction with the other Western Electric employees. Being on a large boat, with more than six hundred people, somehow made people share their common bond of being Western Electric employees.

But it really wasn't until well into the tour on Saturday, till we all went to the same hotel, till we ate lunch in the same dining room, and finally till we took the same disastrous tour, that a real bond was forged. Through the adversity of our trip, our being Western Electric employees brought us closer, to each other and to the company. A rather unexciting, uneventful weekend had been transformed into a very special time. For the first time Eileen felt like part of Western Electric. Noncompany spouses fully became part of the company family.

Social occasions inside and outside the plant, no matter how informal or how ritualized, have a way of effecting a certain amount of camaraderie among workers. Some events, probably once full of meaning, have deteriorated into highly ritualized affairs. They lack all the internal workings that make shared experiences successful. The annual picnic was such an event. At one time it might have been an occasion for management and workers to share an afternoon together, to dispense with some of their work-related differences. It has fallen into a pattern in which the only thing that is shared is the discount to the park.

On the other hand, there is a whole constellation of social events that can be genuinely pleasurable by allowing people to share experiences and emotions. For example, being part of a company-connected trip can change the nature of the outing. The shared affiliation supplies a base for people to relate to an unknown environment and to each other. Such trips are experiences to look forward to, and then to share once one returns.

Banquets, Christmas parties, retirement dinners, and going-away parties are more formalized but still allow people to share generally warm emotions with each other. The going-away parties and retirement dinners in particular are events where friends, whose relationship may be limited to within the plant, are given a last opportunity to mark the importance of the relationship.

16. THE LAST STAGE

THE LAST STAGE—I wanted to interview a person who had a long service record and was about to retire. There was no such person in either of the departments where I worked, so I asked the company's public relations office for the names of some people who were about to retire. They gave me a list of five names. I narrowed the list down to three and decided to interview each of the three and then choose one to focus on. I found them all interesting and different and couldn't choose among them, so I didn't.

Hazel Lefebvre, who is sixty-two, Alba Bocuzzo, fifty-three and Helen Rutyna, sixty-two have 88 years and ten months of service at Western Electric among the three of them. Each was to retire in December 1974.

Western Electric first moved to the Merrimack Valley in August 1943. As soon as it began its operations in a converted shoe warehouse in Haverhill, known as the Haverhill Shops, the company began advertising for help in the local papers, offering pay of fifty cents an hour and no bonus.

All three women began working for the company in its first year of operation in Haverhill. Hazel Lefebvre remembers seeing an ad in the local paper. "They wanted someone who knew algebra. I always loved algebra in school. I had gone to Macintosh Business College, then when I was first married I worked in an office. I did bookkeeping, typing. I worked a year and a half. Then I got pregnant and so I stopped.

"I didn't go back to work until my children were grown up. They were all in school in 1943, so I took a part-time job as a clerk in a store. I worked there about a year and then I saw the ad in the paper.

"By then the children were getting bigger (thirteen, twelve, and nine) and my husband, who worked in the shipyard, was sick a lot so I wanted to get a steady job. So I answered the ad and came in as a tester on the midnight shift."

Alba Bocuzzo was twenty-two when she began working at Western in 1944. She had married Bucky Bocuzzo a year and a half earlier while he was in the service. When he went overseas, she went to work at Western.

Bucky and Alba Bocuzzo, 1943

Hazel Lefebvre and husband, 1955

"I worked at odd jobs before. I was following Bucky when he was switched around in the service. I was a typist up in Vermont. I worked for Ethan Allen (furniture company) and I worked in a boxboard factory. No real good jobs.

"I came home to Haverhill when Bucky went overseas. Western was just starting up, they were offering fifty cents an hour, and I just figured I would go down and apply. They came in August and I think I started in October, so I was one of the first. Right away I liked my job. I got in on the testing end of it. It wasn't crystals, of course, it was something else then, but I can't remember the name of it.

"We were making peanuts compared to what they were making in the shoeshops, but my parents had worked in the shoeshops and they didn't want us children to work there. They wanted something easier for us. My father always had big dreams for me. I was his pet. He'd say, 'I want you be secretary.' He came from Italy and he always spoke broken English. He always told me, 'I want you be secretary.'

"I sort of have a bum arm that I can't use, so I just couldn't do secretarial work too well. I never told him, but he was very happy when I got a job at Western. He thought that was a good kind of job, me being a tester."

Helen Rutyna, 1943

Helen Rutyna began working at Western in January 1944. She had a pretty good job working for Remington company (making $1.40 an hour) in Lowell. "This was just when the war was breaking out, and they had a big plant. I think I worked there a year when they decided to close it down, because of some complication with the government. The rumor was they were selling armor-piercing bullets to Germany. I don't know if that's true, but the plant closed down November 1943 just at the time when the war was in full swing.

"I wanted to go to Western then, but I didn't have a ride, so I applied for unemployment. There were plenty of jobs available but not of my liking. I didn't want to take one of the jobs they offered me in the mills because I had a lung problem.

"They didn't want to give me a check, so I got a note from my doctor. When they were about to give me a check I had gotten a job at Western and had arranged for a ride. I only worked at Western for four months. I lost my ride in April 1944 and had to quit.

"I remember my supervisor, Mr. Schneider. He said, 'Helen, why don't you stay at the hotel? Get a room someplace in Haverhill near the shop.' I said, 'I can't. I've got a mother to support. What'll I have left? I couldn't rent a place, and I couldn't afford a car on fifty cents an hour, no bonus. My six-day-week pay was twenty-six dollars take home, so I had to quit.

"U. S. Rubber was training some of the people from Remington, so I went with about twelve people to Bristol, Rhode Island, to be trained to become a supervisor.

"I was down there for about two months when I got a letter from a girl that worked for Mr. Nordengren. He was a department chief of mine before, and she said there's a fellow that's starting to work at the Western Electric plant from Lowell who could give you a ride, and Mr. Nordengren would like to know if you'd like to come back. I wrote back and told her I would.

"I figured even though I was going to have a good job at U. S. Rubber, it was only a wartime job, and then my boy friend got killed in the service, and everything went down the drain, my plans and everything. I met him at a USO dance and I just knew I cared for him right then and there.

"He felt the same way, so we planned a marriage as soon as he got back. He wanted to get married before he went across but, I don't know, I figured let's wait. My mother said the same thing.

"So I said to myself now I have to look for a job with security, which they offer at Western Electric.

"I'm glad I did. I really like this place. It's been like a home to me."

Those first years in the Haverhill shops were good years for all three women. Helen Rutyna was a tester and soon was holding down a 36 grade job. She loved the work, but her problems with getting a ride to work continued. "All my life I had a problem getting a ride into Haverhill. Finally I bought a car in 1953. I don't think I would have gotten one even then, except my department chief Mr. Ferguson said to me (because I was late a lot), 'Helen, you'd better buy yourself a little jalopy.'

"My ride from Lowell had changed shifts so I was getting rides from people who worked in the shoeshops, and they didn't care if they were late. This one man was supposed to pick me up at six. I'd be standing out on a corner and sometimes he wouldn't show up until seven-thirty. I'd freeze because there'd be no place to warm up.

"I told Mr. Ferguson, 'I'm a nervous type of person, I'd never be able to drive a car.' He told me, 'If you can operate one of these [a test set that took up a full wall], you can operate a car.'

"I'll never forget when I first got my '51 Chevy [she laughs], my mother was my first passenger. I got the car and I didn't have a ride for work on Saturday, we used to work Saturdays, so I drove to work and when I got home my mother was waiting at the gate in her house slippers and her house dress and after finding out how I did, she told me, 'Now you take me for a ride.' I told her, 'I don't know where to take you.' 'Anyplace,' she said. So we went down country roads and everything. When we got home my mother said, 'You're a good driver, you'll be all right.' I've driven ever since, and always owned Chevys.

"I liked working in the Haverhill shops because I liked getting out on my lunch hour and going shopping. I liked going to different restaurants. It broke up my day."

Alba Bocuzzo also liked the longer lunch hours. She says, "We really enjoyed them because you could walk downtown. Oh, it was nice especially at Christmas time. You went downtown and did your shopping during that hour, got something everyday. But over here with only half an hour you're stuck, you can't go shopping or do your errands. Here you've got to worry about hurrying up and getting downtown to do your errands, and of course everything is busy downtown after work, everybody's got the same idea.

"We miss the Haverhill shops really. They didn't have the pressure you have here. In the Haverhill shops you were in big rooms that were not wide open. So you weren't being watched by other supervisors. It was more like a family. You got to know each other well, and you didn't have to worry about what the other side was doing."

When Bucky returned from the service at the end of the war, Alba didn't quit work. "I was doing real well. After a year I was made a 34 grade layout. I figured I'd work until I became pregnant, which never happened, just one of those things. So I kept on working. We wanted a home, so we built one. We built it ourselves. A neighbor told us he had a silo he wanted torn down and if we tore it down we could have the wood to build our house. Bucky tore down the silo and there was a lot of beautiful lumber in there and we started our house.

"We built it practically piece by piece, him and his father. His father was a carpenter, you know—not a real carpenter but he did well for himself. So this is how we built it. It took maybe ten years to complete it, but we lived in the rough for a while—no walls up, just the two-by-fours and rough floor and just the sink and no cabinets and all that stuff.

"We really enjoyed that house. Bucky used to bring his tractors into the living room and put them together there. His friends used the living room too. I mean there was nothing in the living room to worry about, no nice furniture, no nice floor, so everything they wanted to assemble, a boat, a tractor, they did it there. They never wanted to see that living room get done. It was a real playroom."

Hazel Lefebvre, like the two other women, enjoyed working in the Haverhill shops. She did very well. "I came in as a 32 grade tester and a year and a half later I was up to a 37 grade layout operator. In those days they taught you. I started out on the midnight shift, and I went to school during the days. They had a teacher here who taught three days a week. I passed the test and became a 36 grade. And then I got a 37 grade layout operator job.

"While I was in final inspection we used to change places every three months. Mr. Fritz was my boss. He was very nice. My husband used to change shifts every three months in the navy yard, and Mr. Fritz used to help me change shifts when my husband changed.

"I had just moved off the night shift onto days as a 37 grade layout and I got pregnant and had to quit. I stayed out for three years. When I came back I lost my time and had to start as a 32 grade again.

"My pregnancy really surprised me. We got married in 1930 and had three children one, two, three. Then I waited thirteen years before having my other one. I didn't even know I was pregnant. I went to the doctor and he told me. He said, 'What did you think, you were immune?' I said, 'Well, after all these years . . .'

"It was funny. I didn't even give it a thought, not after so many years. I had been working. We had bought a home and were doing good and that's when it happened. I was an old lady then, thirty-four years old. I thought I was old. It's not like it is today. I was really ashamed, I was afraid to tell anybody.

"I stayed out for three years before I went back in. My daughter was three and I went back on days. My husband's mother and father were living with us for a while and they took care of her. My son used to come home from high school and he'd take care of her. Everybody chipped in. When there wasn't anybody home, I'd stay home myself and take care of her.

"Like I said, I went back in as a 32 and I didn't pick up my time for five years. I came back in 1950 and I was laid off in 1950. Things got slow and they had a big layoff. I was only out for five months.

"I worked as a 32 grade for a while and then I got a double upgrade to 34 and then I got a 35, and I was on that for twelve years. I went back on the third shift when a 36 grade job opened in 1961, and I was tired of being a 35 grade. I had been a 37 grade before I had gone out to have my baby and I wanted to get up there again. When I was out for the three years having the baby I lost time. People I started with had those extra three years on me, and so they were passing me and I couldn't seem to catch up and get up on the top of the list for days. So I went on nights to get the 36 grade job.

"I only stayed on nights for six months before I came on days and I've been on days ever since. The first time I worked the third shift years ago I didn't mind it. The children were little and I'd come home and get them off to school and then I'd sleep. But this last time when I went on to get my upgrade I hated it. I hated the hours. I couldn't sleep. I'd go home, sleep about two hours, and that would be it. The children would come home from school and sometimes I'd lay down in the bed but I couldn't seem to sleep and I was always tired. They were building a house right behind us, so when I'd come home I'd hear all that pounding.

"I've been on days ever since for twelve years now in the same department, first upstairs when we set up this department, and then we moved down into the basement about ten years ago.

"I've been offered 37 grade layout jobs, but I didn't care to be a layout operator over here. When I first came to the company and was a layout it was different. The layout operator had more work to do. They were considered almost a supervisor. The supervisor was seldom around, so when you gave work they did it without complaining. But since we've come over here, it's different.

"They don't respect the layout operator over here. They say, 'You're not our boss.' So lots of times the layout has to go to the supervisor to get something done, which in earlier years we didn't. People just don't respect the layouts. I'm surprised because I like to respect anybody over me at all. I always wanted to please people over me. So I'm glad I didn't take any of the promotions, and stayed as a 36 grade tester, because I love my work."

Helen Rutyna kept driving her car to work. When the first Chevy went on her she got another one. "I stayed on the first shift for fifteen years," she said. "I remember I stopped working on the first shift in 1960. I was a 36 grade tester and they offered me a 37 grade layout operator job on the second shift.

"I took it because there was no one to hold me at home. There was no one to worry about or cook meals for. My mom died in 1959. After she died it was lonesome at home. When she was alive I used to rush home and talk with her and fix her meals.

"I fixed her meals the last couple of years because she went blind. I wasn't aware of it until I kept seeing her dropping everything, and stumbling. I took her to an eye doctor. He said she was going blind, and an operation couldn't have cured her because it was hardening of the arteries.

"She went blind in 1957 and then in 1958 she had a stroke. She was bedridden for another year before she passed away, but she was still wonderful company. I'd have a lady take care of her during the day. When I'd step into the house and speak she'd recognize my voice and say something like, 'Oh, my angel's home.' Every child [there were seven girls] was called angel or kitten. I'd say, 'How do you feel, Ma?' She'd say, 'Couldn't feel better.'

"She'd never complain, she'd say she never had it so good, and there she was bedridden, a whole half side paralyzed. She was a beautiful person. I had a good family life.

"I was the fourth oldest of seven girls. My father died in 1940 and he was sixty-two, my age. I took over the responsibility then. My three oldest sisters were married and the others were in school.

"My father worked for the Sacco Lowell shops as a machinist for twenty-six years and they had no pension or nothing. They put money into a pension fund but God knows . . . they moved and he got nothing. I think that broke his heart. He worried himself sick. He had bought the house during the Depression and lost his job. He knew no other trade. He went to work for the WPA, caught a cold on the job, got pneumonia, and died.

"Before my father died he'd take me every Saturday morning, this was a ritual for him, to pay his insurance, pay the mortgage on the house, pay the fuel bill, the electric bill, and a couple of other things like that. He'd show me all the places to go to and from then on that was my job every Saturday morning.

"I still do it to this day. This reporter for the Lowell *Sun* wrote one time, 'I wonder where Helen Rutyna goes every Saturday morning?' I wouldn't dare tell him, he'd laugh at me. But I'm glad I took on that responsibility. When my dad died, he couldn't talk to me, he just held my hand tight. I know he wanted to say take over, and I did.

"Anyway my ma was very good company all those years. When she died it was very lonesome. I'm glad I took that upgrade on the second shift. It was a new job. We were very busy at the time because they were hiring a lot of new people. Most of my people were brand new, not from other departments, but right off the street. I had to train them all. I used to go home exhausted and I'd fall asleep. One day and the other, one day and the other. It helped me through the humps and it was very good for me. I've stayed on that job ever since. I've enjoyed the shift and I really like being a layout."

Alba Bocuzzo was the last of the three women to move over to the North Andover plant. Her group didn't move over from the Haverhill shops until 1960. She worked doing soldering and brazing in the main building for a year before going into an expanding crystal department.

"Ten years ago we moved to the crystal building and I've always loved it. It is one of those places. Everybody who goes over there loves it, and doesn't want to go back to the main shop. It's like the Haverhill shops. It's small. You get to know each other. We have our own cafeteria and a few of us have eaten together every day for ten years. We eat together and the men play cards during lunch. Everyone knows each other.

"I think everybody knows they're stuck with each other and everybody gets along with each other pretty well. There's not much friction. You don't have the boss looking down on you. Well, they do watch you to see that you get your job done, they do that, but they know we're an older group, that we have more experience and they don't need to watch you as closely.

"They know we won't goof off. Of course we've had people in our department who are goof-offs. Some have gotten transferred, and some have gotten upgrades [she laughs]. I mean it didn't make any difference that they goofed off. If it was time for an upgrade, they got it. This gripes you, but what do you do about it–nothing. This is the way the rules read.

"Like I said, I've liked the groups I've worked with and I've liked the work. It's always been interesting. I was never stuck on a bench where I just did the same job eight hours a day. I never did that. I've always fallen into a job where I've changed around to do different operations. I was fortunate that way. That's probably why I stayed at Western. I think if I stayed on one job eight hours a day, it would drive me out of my mind.

"It's never been boring, and I think that's why I stayed such a long time. I never thought I would. It's been a challenge, especially working with crystal. I think I've solved a lot of the problems on my own job. I think the biggest challenge I had was something I solved a couple of years ago and I got a thousand-dollar award from the company. It was a problem which you might not understand because you don't know crystal and it's hard to describe unless I show you the actual thing.

"We were working for flatness on polish plates. Now to look at a plate on a bench here you might go, 'What's wrong with it, it looks flat.' But you have to read it on a special kind of light and it will show you the flatness.

"I worked on this one problem of trying to get the plate flat for nearly three years. We just solved it a year ago. At first I was making plates for the engineer. The plates weren't flat, but we were shipping them because he said they were within limits. Yet when they came over to the supervisors over here in the main shop, they used to complain that the crystals just weren't working properly and they weren't as flat as they should be.

"I knew they weren't flat. I tried to tell the engineer but he wouldn't listen to me. He told me they were within specifications. See, he goes by what's on paper—this should work and this is supposed to work. But I work with the material and I know what works and what doesn't work.

"We were supposed to be working together but he never could understand what I was trying to put across. I'd keep trying to tell him this is wrong and that's wrong. He used to get aggravated and made me feel I was bothering him with minor things that didn't mean anything. It got so after a while I didn't bother telling him anymore. I just tried to work out the problem myself, which I finally did, and if I had any problems I went to my supervisor. My supervisor got real mad at the engineer. I can't tell you the names he used to call him.

"Anyway I worked with my boss until we solved the problem. The engineer didn't want to give me any credit. He gave me a hard time. In fact he tried to pay me off with $185, but my supervisor made a big stink and finally I was given one thousand dollars for my suggestion.

"After my suggestion went through there weren't any complaints anymore from the main shop about the crystals we were sending them. Still the engineer didn't want to give me the credit. Everyone around me, however, knew what I had gone through for maybe a couple of years, struggling with that job everyday, coming in and saying, 'What'll I try today?'

"It was a big challenge and this is what made the job interesting, even though I'd go home with a headache and come in with a headache many mornings. Sometimes I'd wake up in the night and wonder what'll I do tomorrow? Solving that problem was a big thing, a big challenge, and I feel as though I've accomplished something."

Except for a short period when she was a layout, Hazel Lefebvre has been a tester. "I loved it all the time, I loved testing, adjusting—we have to adjust different parts so they read right on the test sets, and if they don't read right, we have to make them read right. You have to find out what is wrong and fix it. It's a real challenge.

"That's why I've always preferred process testing to final testing. I was in final testing for a while and all you got to do was tell if boards were good or bad, you couldn't do anything about it. I got switched back into process as quick as I could because if you found something wrong there you could make them read good if they read wrong.

"It's like solving a puzzle, and I loved that. That's why I don't like these new computers they have. Right now they break down a lot. I suppose once they get the bugs out of them they'll be very good because they can put low-grade people on them because there's not much to do but watch them. You put the program in and that's it. Working on them is like being a final tester and I don't like that. Just sit there and push something in all day long, that would be terrible. I like to be able to fix it myself.

"I'm very proud of what I've done and I've learned a lot. There have been some special jobs that I've been asked to work on. We had some coils at one time that they had to get on an airplane that night, and I did them. Boy, did I work that day. I was testing them and testing them, and getting them ready. We got them out and they thanked me later. And then we had the cable job, going across under the ocean, and I was the only one that worked on the cable. I know they appreciated what I did.

"That means a lot when they can appreciate what you do. I think that's why I liked one supervisor we had better than the others, because he knew how to test. We could work with him, and he would appreciate what we did, and we didn't need to go to the engineers.

"Since then we've had supervisors who don't know how to test. All they can judge you by is how much work you put out, not by how good a tester you are, and that hurts. Like now we got a new boss, we've had him about three months and we're getting another one. We keep changing every four or five months and they don't know what you can do. Other than the amount of work, they wouldn't know if you were a good tester, a bad tester, or what.

"Since we've been getting supervisors who don't know how to test, we have to go to the engineers, and we've been very lucky to have two really nice engineers. It is very unusual. They don't mind us calling them when we need them. They have been real helpful.

"In our department I'm usually busy all the time and I like that. I can't stand sitting around. I like to work every minute I'm here. Once in a while I like to goof off, or have someone around to talk to, but not very often. In fact, now it's getting terrible because I feel that now that I'm retiring I should give up the work to the others instead of doing it myself, so I hang around a bit more than I used to and I'm not used to it.

"Right now I'm teaching people how to do my work. Really it makes me kind of mad to think that I've got to give up my knowledge. All these years doing these jobs, I've got all this knowledge. Lately I've been giving it away.

"I'll leave my notes for the new testers. See, coil filters or other things come in, sometimes the engineers set them up, but plenty of times they write the blueprint up wrong, they give us wrong instructions, so we have to find a way of making them come out right. So that's experience and knowledge. I have all these notes in a little black book. Sometimes I make notes on the blueprints, but there's a lot I've written down in my notebooks over the years.

"Of course parts of my notebooks have been copied, because I use them in teaching my girls and they've been copying them. But there are certain jobs that I'm the only one who's worked on. So no one else has experience on them yet, and if they don't come in before I leave I'll have to leave my notes for a certain way of doing them that I have found works."

Helen Rutyna has liked her work and the company, which she says "has been like a big family. I've been in one department for thirty years and I've had loads of people working with me and they're spread through the whole plant by this time. So no matter where you go, you always see somebody you know and everybody says, 'Hi,' and 'How are you?'

"Some of them now, when I see them out in the parking lot going home from the first shift and they haven't seen me for a long time, they say, 'Hi, Helen,' and they give me a big hug and a kiss like a lost friend. It's nice. I feel at home here.

"They treat you like a human being in here. It may not look like they do, they're a big outfit, and it may look like they're not partial to your problems, but they are. I think so. And I know I'm not the only one. I really like this company, and a girl and myself, we always talk about it. She says, 'Helen, I feel the same way you do about this company, it's been our bread and butter for thirty years.'

"These kids today, they haven't worked in places like we have before, and they've been treated decent and they don't know anything different. But we do. I worked in the mills when I was a teen-ager and they didn't treat you with any respect. I hated it. One supervisor would whistle at you when he wanted to talk to you. I wouldn't answer him. One day he came over and said, 'Didn't you hear me call you?'

"I told him, 'You didn't call me, you whistled at me.' 'Yeah, that's the way I call you,' he said. I told him, 'I have a name, please call me by my name.' He didn't like me since. I was glad to get out of there.

"This place has been my bread and butter. I got up on my feet with it, almost paid my house, took good care of my mom, had the house fixed. When I sit down and take inventory, I've done all that all on my own and I wouldn't have done it anywhere else but here. I had a limited education. I really didn't have a chance to go to school outside of high school, even though I wanted to. No money.

"It's not just the company. People here have been really nice to me. A couple of times on my birthday the girls have gotten together and given me a cake as a little gift to show their appreciation. When my mom was sick they were very nice.

"When my mom died, my house was broken into the day before the funeral and I had cash and bonds for funeral expenses in the house [$420] and it was all stolen.

"The insurance company gave me a hundred dollars, that was all. When they found out about it at work all the kids here took up a collection and gave me a hundred dollars. I thought that was wonderful.

"I've had good jobs since I've worked at Western. I was a tester for a long time and I like it because it was a challenge, I've liked being a layout too. It's a good job, but it's a little hard to please everybody.

"Some girls always think you're giving other girls the better work, no matter how fair you try to be. Sometimes there's one girl who can perform a job better than another one, and when I have to ship my work out, when I have a deadline to meet, I can't give it to the girl who's going to take two hours longer than another girl. I give it to the quicker girl and the first poor girl may feel abused or that I'm mistreating her. I don't know.

"I try to be fair. Some of them understand but it isn't a job where you can make people happy all the time. One day they'll be happy and be nice and everything and the next . . . Women are funny, cranky one day and friendly the next . . . You have to understand, most of the women here are married and have families. Sometimes they come to work and their minds are not on work, but on their family problems. So one day you give them a job and they won't mind it and another day it will hit them the wrong way.

"I try to do the best I can. On a good day when I can go home and there's been no discord I have a good feeling. Other days when things are rough and people are unhappy I go home unhappy because I don't like discord.

"My supervisor has told me over and over not to take it home with me, but I can't help it. I've had to learn the hard way. I have a problem with myself and I know it. I'm a very sensitive person and I get hurt very easily and I shouldn't, not in this type of work.

"I've got an ulcer now. I know that I got the ulcer because I was a worrywort. I know that. When I have some kind of job to do, I worry about it because I want to get it out, and tension and everything like that builds up.

"Well, when I had that ulcer I figured maybe my health is going to give me trouble and I don't want to be staying out two months or three months. I don't want to bleed the company. If I got sick again and had to stay out another length of time, I wouldn't feel right. So I felt that as long as I'm still able, maybe I should get out."

Hazel Lefebvre will retire at the end of December. "I'm sixty-two now," she says. "At the end of December when I retire I'll have twenty-seven years and four months of service. I figure I've worked long enough. I want some time out. I never had time out, really. I'm financially OK right now, so I might as well get out and enjoy a few years.

"I talked to a few of my friends about it, they told me I should get out. They said, 'Why don't you? You're sixty-two, you'll get almost as much now as you would if you stayed to sixty-five. Why wait those three years?' I just decided they were right.

"Then there are little things. I hate to brush the snow off my car in the morning. I couldn't see another winter getting up and brushing that snow off my car. Also I'd like to get up later. I'm so used to getting up early that even on weekends I'm up by seven. However, at the end of a two- or three-week vacation I'm getting up later.

"I want more time with my children. My husband died in 1967, seven years ago, of cancer of the lungs. We were married for thirty-seven years. Welding and everything in the navy yard, that helped cause his death. He had emphysema. I would always be rushing him off to the hospital at night. He wouldn't be breathing. At the end I had to have an oxygen tank in the house.

"The company helped a lot when my husband was sick. The benefits were wonderful. I would have lost my house and everything else if it wasn't for Blue Cross. One hospital bill was $6800. That was paid up, every cent of it. I would have gone bankrupt. My husband was out three years on disability before he died. He was in and out of the hospital all the time.

"They were good with my husband and they give good benefits, but sometimes they act like they don't care. Just two weeks ago I was out sick and they didn't pay me for it after twenty-seven years because I had been out more than five times in a year and I didn't bring in a doctor's certificate. When you're out for a day you don't feel like going to a doctor, you know. What can you tell him, for a doctor's certificate, you just don't feel good? I didn't get paid. If I was in my old department I would have got paid, but we had a new department chief. Sometimes it works out differently. A few years ago I had a small operation and I was out for two weeks, and the doctor told me I'd have to stay out for another week. The company nurse told me I'd have to go back to work the next week and I said, 'No, my doctor says no.' She said, 'They won't pay you,' and I told her, 'Don't, I'm not coming in. If they don't want to pay me, they don't have to.' I stayed out. I got paid."

Alba Bocuzzo is only fifty-three, but she has been thinking about retirement for more than a year. "About a year and a half ago I said when I got my thirty years, anything can happen. If something interesting comes up or if I find another line of work, which I'd like to go into, maybe get a part-time job, I thought I might retire.

"This was in my own mind. I got my thirty years in June. I thought, 'Well, I'm going to get my pension. I'm going to have money

coming in, what do I have to worry about? I have no children to worry about and my husband gets a good pension from the navy yard.'

"In July we went to visit my sister and brother. They both live in Florida. For six years my sister has been trying to talk us into moving down there when we retired. We said we'd look. We took three weeks and took the car and went down one coast and through the middle and down the other coast looking over the different areas for a part that really appealed to us, because my husband is a country lover, he loves trees and grass and hills and that sort of thing, and in Florida you don't get that sort of stuff.

"We didn't find anything and my sister took us around to a couple of places near where she lives, but we didn't see anything we liked. The places were either near the highway or very noisy or very expensive.

"We were all very discouraged. We went to one last realtor. She showed us what she had, three or four places and they had no land, no backyard, which wasn't our style. She said she had one more place which might be suitable but it needed work. She showed it to us and it had a nice piece of land. It was a corner lot, and was just the right size for us, and I said, 'Well, this is our style, what do you think, old man?' I call Bucky old man. And he didn't say no right away so my sister and I worked on him.

"There was something there that appealed to him. He liked the corner and the piece of land. There were trees on it and all it needed was a new roof and cleaning up, which was a minor thing.

"We made an offer and the realtor called us that night and said, OK, it had been accepted, come down in the morning and sign the papers.

"We signed the papers the next morning and left that afternoon. That was it, we were houseowners and we didn't even have time to think about it. All the way home I was really excited about the new house, and moving to Florida and retiring. I couldn't wait to tell my boss, 'I'm retiring.'

"He made me a layout operator a year ago and he said, 'You're not going to leave me for a couple of years now, are you, Alba?' I said, 'No, I think I'll be good for a couple of years.' I thought at least a couple. So I came back in July and told him, 'Joe, I bought a house down in Florida.'

"He couldn't believe that I was leaving so soon. And I couldn't either. I still can't believe it. I won't until I'm there. So much is happening now, between all the moving and trying to wind things up so we can move in January. It's quite a big job, a big step to take.

"I thought of leaving in October. I went over to see Mrs. Nimmo (in the benefits office) and she said if you leave in October you lose quite a bit. I'd lose my five weeks of vacation. She said, 'If you stay until December 16 you have nine days' vacation left for this year. If you save those, you can quit December 16 and get paid through February 6.' So how could I refuse all that for just two months' work?

"I'll be getting about $264 a month, which isn't bad for staying home. It's like $65 a week, and there are cost-of-living raises. If I ever get bored I can always take a part-time job. I'm hoping to find a lot of things to do to keep me busy without going to work for a while. I haven't made up my mind about anything special, just something different that might interest me. I don't know what. I'm just hoping I'll find something in Florida.

"My life has been a lot of routine at work and home. I want to do something different before I get too old to enjoy it.

"I'd like to enjoy a yard, which I never did. I like the sunshine and puttering around the yard, which I never do because I just don't have time. So this is what I'd like to be able to do.

"Bucky opened a secondhand furniture business when he retired from the navy yard. He always loved a store. When I met him he worked in a little hardware store. He's getting a little older and the work with furniture is a little heavy. So we may go into something lighter like bicycles or something when we move to Florida.

"Now we're just trying to get ready to go. We just sold the house and most of our furniture. We're going to take a few pieces, whatever fits in the truck and maybe a small trailer. But we're selling the majority of it. We're having no problems, everything is going along well, except my husband is an awful junk collector, a saver, never threw anything away."

Helen Rutyna is not as excited about her retirement.

"I'm pretty mixed up now that I'm going. I've been here for more than thirty years. I'm going to be pretty lonely. When I work as a layout, I feel I'm needed. A lot of times some of the girls come to me

and confide in me with their problems. They ask for advice and you feel like you're needed and it's a good feeling. I'm going to miss that, because at home, nobody wants my ideas or help or anything. I think if I stay home I'm going to make arrangements with my doctor so I can go to the hospital and do volunteer work at least one day a week.

"I've got to get myself into other projects. I'm going to take a few crafts and hobbies up. I don't know yet. I haven't got a set pattern. I don't think I'm going to get another job. If I were going to get another job, I'd be doing a stupid thing—I'd rather work here. I want to do a lot of reading and tend to my garden and take a little trip here and there and give myself a little time. I have cousins in Europe and some in Poland. I'd like to see them if I could see my way clear with this inflation. I'm afraid to spend the money that I might have to use later.

"I'm going to get a pretty good pension, something like $336 a month from Western Electric alone and $250 from social security. I figured my budget out and I'll have to have I figure close to $600 a month because I still owe on my house and with the fuel bills doubled and the electric bills doubled and the gas that I've got doubled and food, I need something like $400 just to run my house and all the basic necessities. The rest will have to be for emergencies like if the car breaks down. I've just finished paying for a 1972 Chevy and it's going to have to last me a long time.

"I don't have much figured for extras, but I have enough clothes. I don't need any more and I had my house repaired pretty well before I thought of retiring. It's an old house. I bought it from my mom in 1948. I've got a roof over my head and heat and I've got a piece of land to plant my trees and shrubs and whatnot. I've made a real mess of it [she laughs] but it keeps me going.

"I just hope I won't be too lonely. I don't know, this summer the third week of vacation I was getting restless. I had plenty to do around the house. I don't know why it bothered me, but I wanted to go back to work. It's funny, but when I had to come back to work, I found myself singing that day. I felt I had someplace to go and I just didn't feel bad, like some people do when they come back from vacation. I was glad to be back with people, and to have people to talk to."

Hazel Lefebvre also had some doubts about her decision to retire. "When I decided in the summer to retire I was scared. I don't know why but I didn't know whether I was doing the right thing or

not. But now that I am about to retire I'm pretty happy about it.

"You get used to a routine. But I think I'll be all right. I have the children in Connecticut and here. I have grandchildren to visit. I love to drive.

"I'll have plenty of time. I'm going to do some more work for the Pioneers [an association of Western Electric people with more than twenty years' service]. I've worked for them off and on. I was on the knitting committee and on the hospital committee off and on, not too much, a little. But I plan to do more volunteer work. I'd like to use my car to drive people places. I like to drive.

"My girl friend and I hope to go traveling. I've known her ever since I was married. The four of us were always together. She'd have a baby, then I'd have a baby. Then she'd have another one and I'd have another one. She had three boys and a girl. I had three girls and one boy. We were always good friends. Her husband died of cancer, my husband died of cancer. She's got cancer but I haven't so far—knock on wood. I just had my physical and the doctor found nothing wrong whatsoever.

"I'm going to get a pension of $258. And I have social security. I haven't found out what that's going to be. But I've also a pension from my husband, at the navy yard. We had a couple of annuities when my husband died. I had quite a bit of insurance and we had insurance at the navy yard, so I don't have to worry about money. I don't have to worry about a house anymore. When my husband died in 1967 I moved into an apartment in Groveland, and I lived there about a year and then my son, who lives in Salem in a mobile home park, said, 'Why don't you buy a mobile home up here?' So I did. I bought a mobile home right opposite my son and his wife.

"Still I have a funny feeling about going. The idea of knowing you aren't going to work anymore. That's it. You're at your last . . . Well, you're in your last stage of life and that's it. Somehow it's a way of marking . . . I don't know, but you're in your last stage. I'm trying to forget about it now."

17. A RETIREMENT PARTY

About ninety people paid eight dollars apiece to attend the family-style roast beef dinner and retirement party for Alba Bocuzzo at the Rendezvous Restaurant. Alba and her husband, Bucky, sat at the front table along with the three women who had organized the dinner, the master of ceremonies, the vice-president of the local union, and his wife.

Everyone else was scattered off to either side of the wooden dance floor at round tables, listening to the music of a three-piece band. The leader of the trio, Elmore Prescott, a seventy-four-year-old Western Electric retiree, joked with many of his old friends. He told one woman he had just gotten his monthly blood earlier in the day so she shouldn't lean over too far.

Once dinner was finished, the master of ceremonies, wearing a white sports coat, nervously began the evening's speeches and presentations with a series of "spontaneous" jokes, read from notes on an unused computer printout, including the following: ". . . We had a streaker on the schedule, but the police got him. Don't worry, they couldn't pin anything on him."

The jokes and introductions finished, the presentations began with Joe Sifferlen, Alba's boss, talking about Alba and what she had contributed to the company. Then came the gifts. The company's retirement gift, a chiming mantle clock, had been given to Alba earlier in the week. Since Alba had worked in the crystal rooms, she received an engraved desk set with a small uncut crystal mounted on the front. Then she was given a memory book by Millie Sirome—reminiscent of high school yearbooks—in which all the workers in her area had written farewell messages.

Frank Talarico, the union's representative, presented Alba with an honorary gold-plated lifetime union membership card. Later she received a life membership card from a representative of the Pioneers organization.

Alba smiled as she received each gift and heard each nice speech about her, but her face gave hint to the emotions welling up inside. She held them back for a long time, by discreetly dabbing a wet eye.

She could not hide her emotions after the next two gifts, presented by her two friends Eleanor Coburn and Alice Howard. First Eleanor gave Alba the sunshine gift—a leather billfold—and then Alice presented the friendship gift—a leather handbag. Each presentation included a hug, a kiss, and softly spoken words between friends who had worked together for many, many years. By the time Alba sat down again her eyes were much more moist. Bucky sat quietly through the speeches and presentations, supportive, smiling, letting himself be the good-natured butt of several jokes.

Alba and Bucky then stood up and moved to the front of the table for a receiving line. The dinner dishes were cleared away and the band began to play, but the party remained fairly subdued. There was some dancing, but not much. Except for the Hully Gully, when the floor was packed with lines of people moving together in mysterious ways, the dance floor remained fairly empty. Some people began wandering off to the bar.

There were complaints that the dancing hadn't started because the restaurant was having another party in a nearby room and another band playing rock-and-roll was blaring competing music.

By ten o'clock the party had thinned out. The first to leave talked about needing to get up early for work on Saturday. By 10:30 no excuse was needed, people just left. Bucky and I were sitting in a corner

working on some drinks and long cigars. Alba suggested that my wife Eileen and I come back to her house for a while.

There were less than twenty people left at the party. All of a sudden the party began to pick up. It took on a new life, which seemed to suck all the remaining people into it. There were polkas, Charlestons, rhumbas, and more Hully Gullies. Even my wife, who is usually shy at such affairs, rose with seven or eight other women, first to learn the Hully Gully and then to be led by a waitress in a long, slow line in an attempt to master the steps of a Greek dance.

Several efforts were made to get Bucky and me into the dancing, but we contented ourselves with new cigars and new drinks. Those who remained dancing soon moved to the side of a baby grand piano and began singing old tunes. Off key, on key, right words, wrong words, the tunes kept coming: "I'm Looking Over a Four Leaf Clover," "Oh Marie," "My Wild Irish Rose," and many more.

We were now a group of fewer than ten, but the evening seemed quite young. The trio, Elmore Prescott at the piano, Charlie Nielson on the sax, and Peter Moro on the drums, played as though they could play all night.

It was well after twelve when we again started talking about going back to Alba and Bucky's. But we couldn't make it, and soon the idea was dropped, and the party ended. Slowly, the band packed up its instruments and carted them out into the frosty evening.

A few last remarks, a few last kisses and hugs were exchanged. It had become a helluva party, everyone smilingly agreed. The final touch, Bucky and I agreed, was one more cigar for the drive home.

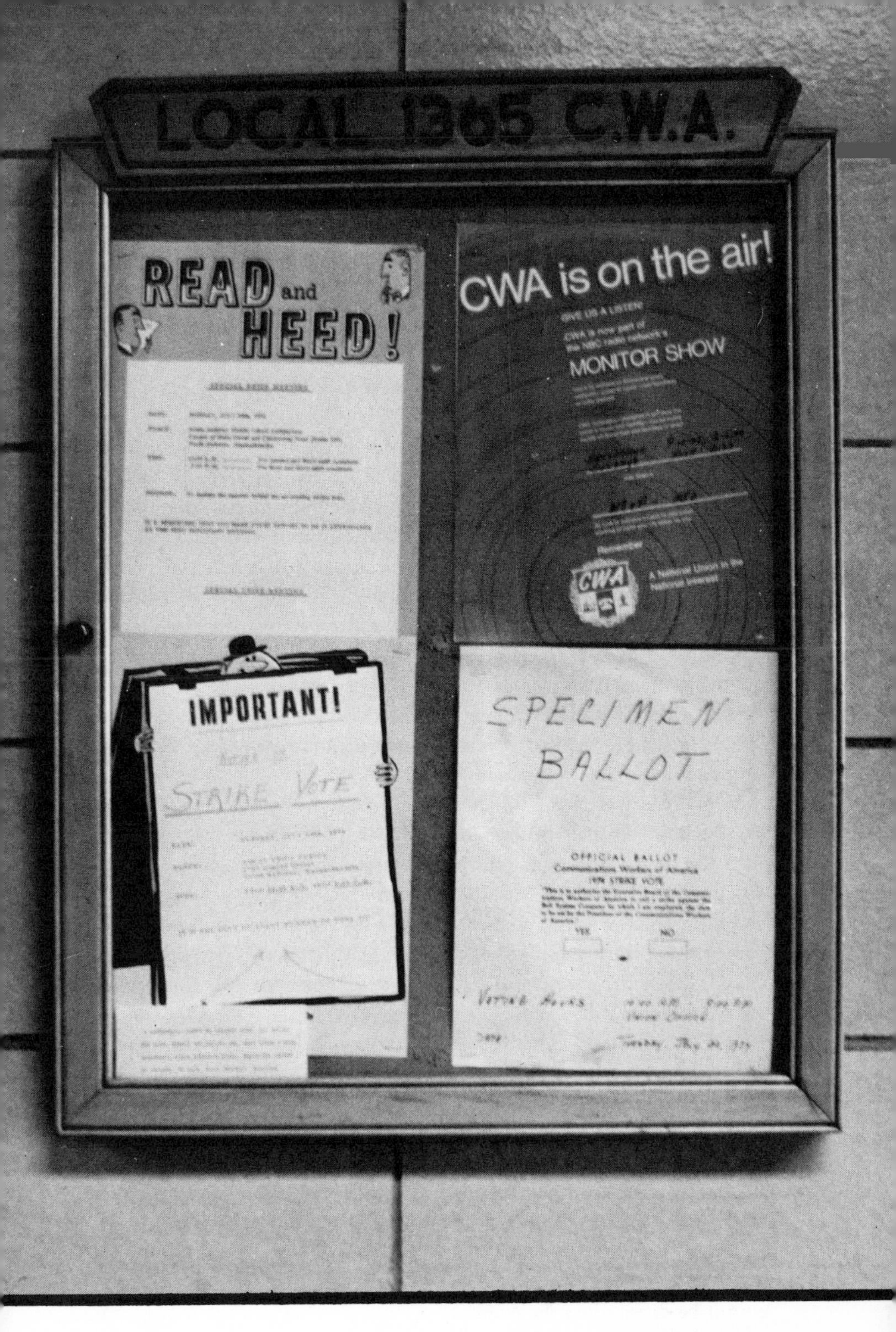

18. LOCAL 1365

During my first day at work a woman in the personnel office told me, "We have a modified agency shop at Western Electric. If you choose to join the union [Communications Workers of America (CWA) Local 1365], your dues can be used as the union sees fit. If you don't join, $1.50 will be deducted each week from your pay and will go for negotiations."[1]

For the first three months on the job I waited for a shop steward to introduce himself and welcome me to Western and the union. When that didn't happen I asked Arnold Thompson[2] and Mario Caluchi, who had started with me if anyone from the union had contacted either of them. Both said no, so we decided to find out who our shop steward was. No one in our department seemed to know, but Ann Hill said she knew a woman in the next area who was a shop steward, and that she'd get some membership cards for us. About a week later Ann brought Arnold, Mario, and me three membership forms. We filled out the forms, returned them to her, and about a month later my union card arrived in the mail.

Very little was ever said about the union. Actually I can remember seeing shop stewards in our area only once. We were having trouble with the rates, and one day a rumor started that the union was coming down to fix things up. A few days later two men, who were reported to be union stewards, came into our shop, sat down, and had a long conversation with Jack, our supervisor. Although they supposedly had talked about the rates, nothing seemed to come of the conversation; the rates weren't changed.

I can recall no other visible union activity. I never heard anyone talk about filing a grievance, or having a grievance handled by a shop steward. Periodically, one of the union's bulletin boards would have a notice about an upcoming union meeting. None of the people I met ever went to any of those monthly meetings. I was curious to know why the union representatives on the shop floor were not more visible and why people didn't talk about the union, so I began investigating.

[1] I later learned from union and company officials that this was untrue, that there are no restrictions attached to the $1.50 a person is required to pay even if he doesn't join the union.

[2] The names in this chapter have been changed.

Mary Sullivan told me, "I've always been in the union. I know things would have been worse in here if there wasn't a union. I can't say I've done much. Really, I've never gone to a meeting, but I'm glad we have a union."

Arlene Dowd told me, "I joined the union the day I came into the shops, but I haven't been to a union meeting except for strike votes since I came to Western twenty years ago."

Most people I talked to generally agreed that the union had helped improve conditions and wages. However, many didn't think that highly of the local, and the current leadership. Several people told me that the union, like the company, had gotten too big and too impersonal. One man, Nunzio Lagotti, told me, "It's not a real strong union, too many guys are hustling—trying to get ahead. A couple of big shots are working for the company." (Two of the local's most recent leaders, including the local's well-respected past president, who had held the post for fourteen years, are now working in management positions.)

I didn't hear much about the union until well after I stopped working, when there was a flurry of activity about contract negotiations in July 1974. The Merrimack Valley Works closes for two weeks in July (13–28) and only maintains a skeleton crew. Just before the summer vacation, rumors started to circulate around the plant that a settlement wouldn't be reached before the current contract expired on July 17 and that a strike would have to be called.

Many people leaving for vacation appeared to be prepared for a strike. "No way it can be avoided," I was told. Chuck Lawrence said, as if he had inside information, "Looks like we're going to have a long vacation." He already had some odd jobs lined up for after the vacation.

Not everyone wanted to go out. Several people told me they hoped a strike could be avoided. Ruth Perry said, "A long strike will really hurt. You never get back the money you lose when you're out of work very long." Listening to people, I realized how economically taxing a strike would be. A long strike could wipe out much of a family's savings, and a short strike could inflict a serious strain on a family budget. Nevertheless, I felt a sense of resignation to the need for a strike. People were concerned about inflation, since increases the company was offering did not seem adequate. When the plant closed on Friday, I expected a strike and only hoped it wouldn't be a long one.

John Gearen thought I was wrong. He didn't think there would be a strike. He was sure it was all a lot of talk, a big show the union and company were putting on. He thought they'd argue and threaten each other up to the last minute, and finally reach an eleventh hour settlement.

For the first time the CWA, which represents both Western Electric and Bell System employees and claims a membership of more than five hundred thousand workers, was going to bargain on a national level. Until this time bargaining had been handled on the local level.

During the two weeks the plant was out, there were occasional reports in the papers on the progress, or the lack of progress, in the bargaining. July 17 came and went and a new contract wasn't signed. Both the union and management agreed to a day-to-day extension of the existing contract. The CWA's international president said that the company's last offer was "inadequate." The union was hoping the company would agree to a full agency shop.[3] As for wages, he said they couldn't be expected to negotiate for less than 10 per cent if they intended to keep pace with inflation and productivity. He told a news conference that his union was seeking an annual increase of about 14 per cent.

The company disputed the union's claim that its money offer wasn't adequate. An A. T. & T. vice-president and company spokesman called the offer "very substantial" and said he was "disappointed an agreement wasn't reached."

When the plant reopened on July 29, people had only the vaguest idea of what was happening with the bargaining. Always before they had gone to local officials for such information, but this time they had little to report. Negotiations were stalled, and union officials had called for a national strike vote. At our location the vote was to be taken the next day, July 30.

[3] Prior to 1971 there had been no type of agency shop at the Merrimack Valley Works. Local 1365, in its nearly twenty-five years of organizing, had never been able to organize more than about 55 per cent of the possible bargaining unit. In 1971, during negotiations, the company agreed to a modified agency shop, under which all new workers who entered the company's employment after the contract (1971) had to join the union or pay union dues. Workers who were on the rolls before 1971 and didn't want to join the union weren't required to. So just before the new bargaining began the union was representing 5,350 of a possible 6,600 workers. Under the terms of a full agency shop, all workers in the bargaining unit even those previously not organized, would have to join the union or pay union dues beginning January 1, 1976.

That night a meeting was held at a nearby middle school. A similar meeting had been held at ten in the morning for people working on the second and third shifts. John Keker, president of the local, and other union officials stood at the door. You had to show your union card to get into the meeting. The press wasn't allowed in.

Usually fewer than forty people attend union meetings. More than one thousand people were present at the meeting that evening. Local officials, the bargaining committee, the International's representative, and about thirty shop stewards sat on the stage.

John Keker chaired the meeting, but Mike Luccino, the International's local representative and another past president of the local, was the person who spoke about the negotiations. Luccino explained that negotiations weren't going well. He said the International was still dissatisfied with the company's offer. The first year's wage increase was not high enough, nor were the following two years'. He asserted that new coverage, such as dental care, which the company had proposed to begin in 1976, should be moved up.

Luccino told us the International hoped a strike wasn't necessary, but that they needed the backing of the people in case a strike had to be called. We were told how similar votes were being taken at locals all around the country. He urged us to turn out in large numbers and to give the local an overwhelming strike vote. He then told us that the voting hours would be from 10:00 A.M. to 9:00 P.M. the next day at union headquarters.

A fellow in the front of the auditorium raised his hand, was called upon, and asked who had set the hours. Luccino said local officials. The guy said how come the hours had been set like that and why couldn't the hours start earlier, so people could vote on their way into work. Luccino said the times had been set and that was all there was to it, that they couldn't be changed.

Now people in different parts of the auditorium got up and started asking why they couldn't be changed. People began booing Luccino's answers. Holding his position, Luccino said first that since these hours had already been posted they had to be stuck to, and then insisted that there weren't enough people to man the desks for the earlier hours.

One man in the back stood up and said, "Who the hell is this union run for anyway? Can't you see we want the hours earlier? Why don't you change them?"

Luccino was getting madder and repeated that the hours couldn't be changed. Another man got up and said, "Hell, when there are union elections you open up early. Why can't you do it when we want it?"

The meeting, which had been going well, was getting ugly. John Keker and Mike Luccino conferred for a moment and then Keker said that they'd change the hours and open up early. This announcement was greeted by long and loud applause, and several people volunteered to man the desks at the earlier times.

The meeting soon ended and people began filing out of the auditorium. The man I had been sitting with was an old friend of Luccino's. I told him that I thought Luccino had handled the situation very badly. He said he couldn't understand it either. "Mike's one of the best people this union's ever had," he told me. "I can't understand it, it's not like him. He just lost his cool."

The next day more than four thousand of the fifty-four hundred workers in the local bargaining unit cast votes. Although the ballots weren't counted locally, it was obvious that an overwhelming strike vote had been given the union. Judging from the meeting and all that had been said, people felt it was only a matter of days or weeks before negotiations would break down and a strike would be called.

The people at the Merrimack Valley Works have struck before. The CWA first organized workers locally in 1944, less than a year after Western Electric had opened the Haverhill shops. Bob Lamont, who has twenty-eight years of service with the company, remembers those times. He was one of the first to join the union and has held office in the union or been on its executive board for twenty-five years.

"Those early days," Bob says, "were tough. Some of the guys who started the local were a little crude and coarse, but they got it started. In those days it was a constant battle to organize the place. We were lucky to have 55 per cent of the people organized. You didn't have to sign up, so it was a constant struggle. After a layoff there would be a dip in membership. There was an escape clause, and a worker could get out of the union. It was tough, but at least the company didn't try to bust the union.

"People up in these parts, especially telephone workers, are very independent. They don't do things just because you tell them to. It isn't like coal miners, and steelworkers. Down there they say, "We're going to strike." Up here you tell them you want a strike vote, and they're going to ask you why?

"We've had to strike a couple of times. We went out in '49 or '50, I can't remember the exact date, for six weeks. We were getting the pants knocked off of us. So we settled. We could have gotten a better settlement before the strike.

"We went back out in 1968 and then again in 1971, over a number of issues. We had to show the company we could pull the people out. The company found out that there weren't going to be any scabs, even though there are a number of people in the plant who aren't in the union. In 1968, the first day of the strike, only thirty-two people crossed the picket line. A lot of them didn't know about the strike the first day. After that there were only one or two people who crossed the line. One guy would hop over the back fence, but there really weren't any scabs. We didn't stay out long on either strike, but it was something we had to do."

The strike vote had been given, but the strike was never called. Within ten days a settlement had been reached. The first day after the announcement had been made, I heard a number of disgruntled remarks about the proposed contract. One woman told me, "They got the agency shop they wanted, and they sold us down the river for it." Another person told me, "They didn't do anything for us. All those promises, all that talk, and we didn't get anything."

John Shattuck, who had heard a lot of the squawking, told me, "I know people think they got sold down the river, but that's because they don't understand things. The union used the people, led them on, so they could get the strike vote. What do you think, they were going to go to the meeting and tell people that they wanted the vote, but weren't going to strike? Hell no, they couldn't do that, but that's how it is. They wanted the strike vote for the bargaining."

Bob Lamont was one of the few people who didn't see it that way.

"I know a lot of people don't like this last contract. They think that the union got the agency shop and didn't get the wage package or benefits they wanted. But it isn't true, it's a damn good contract. It's the best contract we've ever gotten. We got plant wide bumping, which is something we've been trying to get for years. We got a full agency shop which we've been trying to get for fifteen years, and we got good benefits. You'll see, people don't realize it now but they should see how well we did. You can't convey it. People read what they want to read. In a couple of years they'll be able to see just how well we've done."

The hostility I encountered on the shop floor carried over to the ratification meeting, which was held on a Sunday. Fewer people, maybe six hundred, attended the meeting, and local officials tried to explain the proposed settlement. Some bitter and heated exchanges took place during the meeting.

Several people remarked that the contract they were now being asked to ratify, except for the inclusion of the agency shop provision, was very similar to the one they had been told was totally unacceptable before the strike vote.

By the time the meeting ended, more than three hours after it had begun, fewer than half of the people remained in the auditorium. For several weeks after the meeting I kept hearing complaints about the negotiations. Most were focused not at the local but at the International. One person summed up a lot of the ill feelings when she told me, "The International fooled us. They got us all worked up for a strike vote. They made us think we were really going to get something with this national bargaining. Then they pulled the rug out from under us. They got their agency shop, and we didn't get nothin'. You ask me this national bargaining ain't nothin'!"

The shop floor was rife with bitter complaints about the contract and how "the workers" had been dealt a "dirty deal." I was somewhat surprised in the midst of this to find a number of people who turned at least some of their anger back on themselves. Helene Sanders told me, "It makes me purple when people say we were sold down the river. It's our fault. We don't participate. We don't go to church, we don't vote, and we don't go to union meetings. It's like the police, we don't call them unless there's a rape or a theft. Then if they don't do exactly what we want them to do we complain and yell bloody hell."

Another worker, Jack Melinko, was more critical. He told me, "I was a union steward for a couple of weeks ten, twelve years ago, but I quit. I couldn't take it. The only time people in here bitch is when something happens to them personally. They don't give a shit about anyone else but themselves."

Ruth Davis, a shop steward, told me, "This is the first time in the fifteen years that I've been in the union that I don't agree with what they did. I think they didn't do right. They geared everyone up for a strike and then we took a step backwards. They were wrong, but when I hear these people in here complain it burns my ass. They had the chance, they could have rejected the contract. I did and I can't afford to be out. I know some people think this is a weak union. It isn't but if they think it is, it's because of them. They let it be that way. They want to be spoon fed, and not do anything, and they deserve everything they get."

It took me a long time to figure out how I felt about the union. When talks began with company officials about my coming to work, I was surprised that a union official was not included in the deliberations. Since the company felt confident enough that they could get the union to go along with their approval of my study, I sensed that the local couldn't be that strong.

While working, I encountered virtually no union activity, and what I later saw of the contract negotiations was disappointing. When I told union officials all the negative things people on the shop floor said about the union, they didn't seem surprised.

Don DeLuca, the union vice-president, told me, "We've heard the same complaints. We do all we can. I'll tell you people complain but there isn't another local that's gotten more for its people than this local."

Even so local union officials admitted that they could make improvements, especially in the areas of shop steward selection and training. They talked about being overwhelmed with work, that they could only handle so many cases. They also echoed company officials who would respond to severe criticism workers had about the company by saying, "A lot of people don't see what we do. We don't broadcast the times that we help workers." Bob Lamont said I really couldn't appreciate what the union did until there was a strike, or a layoff. He said, "They don't come to us much unless they have a real need."

What Bob said sounded good, but his words didn't mean much to me until the company laid off approximately a hundred shop employees, because of the recession, and bumped other workers. If a person with long service loses his job, he can bump a person in the same grade with less service. That person then may be downgraded and may bump another person with less service. For the week following the layoff announcement the union office, which is located across the street from the plant, was packed with people seeking advice and assistance, and the phone didn't stop ringing.

Local 1365 is neither a militant local, constantly battling the company, nor is it a company union. By and large it maintains what both sides would call a good working relationship with the emphasis on conciliation wherever possible.

Generally people on the shop floor see the need for the union, but they don't view it as a very active or strident force in handling their day-to-day work-connected problems. Whatever the reason or set of reasons, Local 1365 has evolved into a necessary bureaucracy that deals with an even larger bureaucracy.

19. FRINGE BENEFITS

Why do people choose to work in the shops at Western Electric? Money seems the most logical answer, but there are places in the valley where a person can make more. Other than wages, workers most often mention steadiness of employment, good working conditions, and fringe benefits.

"I got out of the service," David Rosen[1] said, "and I went back to work in the shoeshops. After a while there wasn't enough work, the wages started going down, and they started laying people off. It wasn't for me, not with a wife and two kids, so I came over to Western. I knew it was steady. It's about the steadiest work around these parts. The wages weren't so good to start, but I knew I could get ahead. I can't complain, I'm doing OK."

Jack Murphy told me, ". . . I've got friends over at Raytheon. They may make out a little better than me, but there's always the chance they'll get laid off. One of the guys I used to work with left here to go to work at Raytheon about eight years ago. Well, they lost some government contract, and he wants back in here. He lost all that service. I'm glad I've stayed. I figure people are always going to use phones and need phone equipment. I've got a good steady job."

Western Electric is aware of what attracts workers. The management is quite proud of the working conditions in the plant. It also realizes that the security it offers to employees, in terms of steady employment, is very attractive, especially in a geographic area where economic insecurity—layoffs, seasonal slumps, plant closings—has long been a part of the scene. In the thirty years that Western has been in the area, the plant has never had to close because of economic conditions. Layoffs have been infrequent and have generally been smaller and shorter in duration than at most other manufacturing operations in the area.

In the time I spent in and around the plant, I learned just how important security was to many workers, especially ones who had lived through the Depression as children and had parents out of work or in

[1] The names in this chapter have been changed.

unsteady jobs. For many of them having a secure job was worth making a little less money.

The relative cleanliness of the plant was attractive to many workers, particularly women. Women quite frequently spoke of how much cleaner Western was than the shoeshops and textile mills of the area. Dottie Johnson's remarks to me were typical. She said, "I like Western because it's clean. If you've seen some of these other plants around here, they really are filthy. You have to wear old clothes, and you can't help but get dirty. It's not worth it, not for me. It feels less like a factory in here."

Security and decent working conditions appeal to large numbers of shop employees. Interested in keeping the workers it attracts on a long-term basis, the company offers a complicated system of fringe benefits.

During orientation a new employee is given a folder full of pamphlets, two of them describing in detail the fringe benefits. A girl sitting next to me who had just started work again after a two-year absence said, "They've really improved some of the coverage, especially in regard to maternity benefits. When I worked here before, they didn't pay much. My doctor bills were about a thousand dollars and they had maximum coverage of $320. Now they'll pay up to 85 per cent of your medical bills connected with maternity, up to a thousand dollars. I decided to come back. My husband and I would like to have another kid. This way I'll work for a couple of years and then we can have the kid and it'll be pretty much paid for."

A young man of nineteen, sitting on my other side, told me, "I was just about to get married and I couldn't afford not to come to work at Western. [They pay Blue Cross, Blue Shield, major medical.] In here you're really protected. I could have gotten another job, really a better paying job, at a small outfit, but they didn't have anywhere near the benefits they got here. I had my uncle who's a supervisor put in a good word for me so I could get a job."

The fringe benefits improve with length of service. Even a cursory look at the benefits makes the advantages of staying more than apparent. For example, on Blue Cross-Blue Shield coverage the employee is responsible for paying the full amount for the first six months of his employment. After that the company assumes the cost. A worker needs six months' service to qualify for the Extraordinary Medical Expense (EME) plan and for group life insurance. An employee who might not be particularly happy with his work the first six months will think twice about quitting when the large medical expense is carried by the company.

Of all the benefits, I found medical coverage and pensions were most often discussed. Teddy Smith once told me, "The medical coverage I got here for my family would cost me at least three or four hundred dollars a year if I had to pay for it." Several people told me they would have been ruined if it hadn't been for the medical coverage the company provided. Mary Stack told me, "When my father was sick he ran up hospital bills of $6,800 and it was all taken care of by the company." Warren Wyzokowski told me, "My two daughters both were sick this past year. One needed an operation, and thank God it was all covered by the company's EME plan."

Medical coverage was particularly appreciated by the people who had the occasion to use it. Pensions, on the other hand, touched a larger segment of workers. Once again it was the older workers who most frequently talked about pensions. I heard several bitter stories from people about their parents having worked for companies for twenty or thirty years, and having retired with nothing, and how they didn't want to be caught in the same situation.

I heard a number of people say that they were staying at work primarily for the pension. Rosie Smith told me she didn't really like her job, but then added, "I might as well stay. I've already put in ten years, another five and I'm entitled to a pension."

Not everyone, however, is willing to stay the maximum time for a pension. Daphne Dwyer told me, "I'm only sixty and I've already got twenty-eight years in here. Two more years and I'd have thirty years of service. I thought about staying those two extra years, but I've decided not to. I don't mind my job, but I want some time for my husband and grandchildren. Who knows if I'll still be alive in two years. It means getting less, but it's time for me to get out.

On the other hand Peter Dublin, who's forty-six, had eighteen years of service when he recently quit. He said it wasn't worth waiting around for the pension. I had heard so much about the pension that I decided to find out what it meant in cash terms.

No worker with less than fifteen years of service is entitled to a pension. A worker with fifteen years of service can get a pension only if he retires at sixty-five. An employee needs at least twenty years of service to retire at fifty-five, and twenty-five years to retire at fifty. A worker with thirty years of service is entitled to retire at any time.

Pensions are computed as follows:

1 per cent Average monthly rate of pay for highest five years

×

Years of credited service through month age fifty-five is reached

1.5 per cent Average monthly rate of pay for highest five years

×

Years of credited service after month age fifty-five is reached

I asked the company to compute the pensions for two hypothetical situations. A 32 grade bench worker retiring in June 1974 at age sixty-five with thirty years of service would receive a monthly pension of $221.67. The same worker retiring at age fifty-five and with twenty years of service would receive $126.67 a month.

The figures surprised me. I couldn't imagine holding onto a job for twenty years in order to get $127 a month on retirement. Several younger workers expressed similar sentiments. Peter Weiner told me just before he quit, "Sure they got good benefits, but I'm not going to stay on a job I don't dig for twenty years to get a pension." Pietra Toelle, who is twenty and has worked for the company for two years, said, "I'd like to get out of here, but I'm going to keep working until I finish school. The company pays for my courses. When I finish them I'll look for a better job, maybe something in business."

One day I got into a long conversation with Evelyn Walker, who was about to retire after thirty years, about how I didn't think the benefits would hold me to a job I didn't like, and what young workers I talked to said about the benefits. She surprised me when she said, "I think these kids come to work just looking for a fast buck. They want easy money. I walk through the plant and see all these young people and they don't care whether they work or not, they're very independent. They figure if they get fired they'll go get a job somewhere else. This is the way they feel.

"My generation is different. We were brought up in the Depression and we had to work for our dollar. That's why the security and benefits in here means a lot to me. You might not think the pension is much, but my father retired after thirty-five years and got nothing from the company he worked for, not even a penny. My pension and my husband's plus social security will help make me and John independent."

The company and the workers don't always agree on what constitutes a fringe benefit. The Merrimack Valley Credit Union is not considered a fringe benefit by the company, but it is viewed as one by workers. The credit union was originally organized by the Communications Workers of America for its workers in 1955. It is now run for the benefit of all Merrimack Valley employees. Its current manager, Larry Harper, explained to me that of the approximately 10,000 people employed at Merrimack Valley 93 per cent belong to the credit union. With $17,000,000 in savings it is the hundredth largest credit union in the United States (there are 14,000) and the second largest in the state, behind Polaroid.

Mr. Harper is quick to point out the special features or services, as he likes to call them, of the credit union. "The company has been kind enough to rent the credit union space in the plant at a nominal fee. We are open during working hours for people either to make deposits or withdraw money. We have a telephone service so that if you call up in the morning and want to withdraw some money we'll have it ready that same afternoon. Employees can also sign up for payroll deductions, either to have money go into their savings accounts or to pay off loans.

"We are," Mr. Harper continued, "able to respond to personal needs quite quickly. I'll give you an example. There was a man working in the shops. He found out that his father was dying and he needed to go home to Puerto Rico to be with him. We got him the loan for the plane tickets and everything that same day.

"There are certain limitations we have on making loans," Mr. Harper said. "An employee has to be with the company for a *year* before he can borrow money. The credit union will not allow a person to have more than four loans at any one time, and, with few exceptions, it can't loan more than $10,000 to a person."

Of the nearly 10,000 loans made last year to 6,500 different people, which totaled more than $15,000,000, the two major types were car and personal loans, accounting for 60 per cent of the total loans. There were over 3,000 car loans made ($5,000,000) at an annual interest rate of 9 per cent. There were nearly 3,500 personal (signature) loans made ($4,500,000) at an annual interest rate of 12 per cent.

Most of the people I met used the credit union both for savings and for loans. I would say that better than two thirds of the people I knew had amounts automatically taken out of their weekly checks and put into their savings accounts at the credit union. The amounts varied greatly, but the basic range I saw was between ten dollars and twenty-five dollars a week. A couple of women told me they had as much as forty dollars a week taken out. One woman told me, "When I get my check at the end of the week it really looks small. I think I take home less than forty dollars a week. We save money from my salary, and then we're paying off two loans, one for a car I bought a year ago, and the other for a fence we put up around the house. Sometimes I say something to my husband, that I feel like I'm not making anything, and he says we do much better at the credit union than we could elsewhere."

Many employees avail themselves of the Christmas clubs (which pays regular dividends) and the savings club for trips. Johnny O'Sullivan told me, "I throw a couple of bucks a week into the Christmas club. That way when Christmas comes I've got about a hundred bucks to be a big shot with my wife and kids. If I had to get the money up just before Christmas, forget it, but what the hell is a couple of bucks a week." Dottie Jacklin told me she and a few girl friends regularly put money aside for trips. "We plan them more than a year in advance and then we start saving. It's easier that way. It makes the trip more real. You can see the amount you've saved for the trip growing all the time. Last year we went to Honolulu. I loved it, and I'm ready to go on the next trip. I want to go on a cruise. So three of us have decided to go next January. We've already begun saving."

Many people feel that it is advantageous for the credit union to be in the building. On a particularly hot day during the summer, a woman I was working with said, "I've had it. It's too hot in here and it's too hot in my house. I want an air conditioner." She put down her tools and walked out of the area. She came back a couple of minutes later and told me, "I called my husband at work and told him I wanted an air conditioner. I've wanted one for a few years. I told him I had to have one now. He said OK and told me to take the money out of my savings account at the credit union. I called them, and they said the money would be ready for me this afternoon. Imagine," she said, "this evening I'll have air conditioning."

The rates the credit union charges for loans are, in general, lower than commercial institutions would charge, and the interest they pay on savings (6 per cent) higher, and yet some workers see the credit union as a mixed blessing. One day, coming down from lunch, Frank McDonald said to me, "Look out there in the parking lot. I wonder how many of those cars Western Electric owns. You know," he continued, "I did a dumb thing. I worked here a year, and I bought an expensive car. I wanted it—I never owned my own car, but I had to borrow about four grand from the credit union. It made it so I really can't leave. I couldn't just take off and fool around because I couldn't keep up with the payments without this job. I think maybe if I didn't take that loan, I would already have left here."

Barbara Lacey told me a story that I heard countless times. She said, "I came to work at Western Electric when I was eighteen. I had payments to make on a car so I came here. I said to myself, 'As soon as I pay the car off I'll quit.' Now I've worked here twenty years and I owe more money on my 1973 car than when I first came here to pay off a bill on that other car. I just got more and more things, and I like it. I've gotten used to it. I'm sure when I pay off the '73 I'll get a '75."

Implicit in these remarks is the feeling that the credit union is part of the company, and that one can't leave the company if one has an outstanding loan. The fact of the matter is that the company does not own or control the credit union. It is an independent body that functions in the plant building. As for the worker's ability to leave the company while holding an outstanding loan, the situation is probably no different than if a loan had been taken out from a commercial bank. One can leave Western Electric and still be indebted to the credit union. There was nothing stipulated in any of the loans I saw which required a person to maintain employment at Western. Nevertheless, almost everyone I met felt the credit union was part of Western and that once they had a loan their mobility was dramatically reduced.

Fringe benefits and job privileges undoubtedly reduce the mobility of many workers. If benefits were transferable, and pensions were vested with workers rather than companies, a worker would have greater flexibility in seeking satisfying work.

Under the present fringe benefit and seniority privileges systems a worker will have serious doubts about leaving a job for another company after a few years of service, even if the job is not completely satisfying. Quitting work and starting fresh somewhere else involves many risks. That is too heavy a price for many workers. They choose to remain on jobs that are not completely satisfying in order to hold onto the benefits and privileges they have earned. To a large extent, then, loyalty is bought.

20. ALL IN A FAMILY

IRENE COLLINS—My earliest recollection of Irene is of the kidding she took when she reported she had taken her driving test. The first time we talked together was at the end of inventory week, when a group came in to work Saturday. At the end of the day, four of us, including Irene, went across the street to Jimmy's Place for a pizza. Irene's daughter Joyce joined us. I liked Irene and Joyce, found out that both of their husbands worked at Western, and thought it would be interesting to interview them.

It is not unusual to find more than one family member working in the shops at Western Electric. Pat and Irene Collins work there, as do their daughter Joyce and her husband Danny. One evening all five of us sat around Irene and Pat's kitchen table talking about work and family, the past and the future. I began the conversation by asking them where they had grown up.

Pat was the first to speak up, recalling the noise and arguments in crowded streets when he was very little. "When I was five," Pat says, "me and three sisters went to an institution. They were very strict there. Seven years we were there and then we moved back with my mother and stepfather. I wanted to turn twelve so bad, and you know why? So I could get my shoeshine license. I shined shoes till I was fourteen and sold newspapers.

"Then we moved to Cape May, New Jersey, and then to Clinton, Oklahoma. My stepfather wanted me to finish school and go to the service academy, but I didn't care about marks then. I quit school and came back to Lawrence. I worked in the mills. I was sixteen then and worked thirty-seven and a half hours a week, the limit for a kid my age. I made twenty-one dollars a week. I went into the service in 1944 when I turned eighteen. I served for three years and five months. Geez, I could go on like this all evening.

"When I came out of the service I went to work at Malden Mills. I worked there for six years and when work got slow (1956) I moved over to Western, and I've been working there ever since."

Irene Collins is one of nine children. Her mother died soon after

Irene was born. The older children helped the father keep the family together. Irene finished three years of high school before quitting to begin working in a laundry, making about twenty dollars a week.

"I met Pat at a dance before he went into the service," Irene says. "I didn't like him to start with. He was with my girl friend but we started going out. I felt good in his arms one night and that was it. I couldn't get rid of him."

"Don't get smart," Pat remarks.

Irene laughs and says, "We got married soon after Pat got out of the service and this September it will be twenty-five years together. I went to work at Western the first time around in 1956 when Pat started there."

"No," Pat says. "We were living in the projects then and you went to work on the third shift over at ITT."

"Oh yeah," Irene says, "that's right. I worked over there for six months and then worked over at Western for a couple of years before quitting to have the kids."

Pat interrupts. "The reason you stopped working there the first time was the kids were getting hurt. One cut himself with a beer bottle and this one [Joyce] got a broomstick through her mouth."

"That's right," Irene says. "I quit to be home more and then I started having one kid right after another. I worked off and on for a while, but really I didn't work for nine years. I went back to work six years ago. I started up at Reliance, but things were slow, and at Western it's steady and dependable, and having worked there years ago I decided to go back."

"Danny, my mom, and me, we all started working at Western within a month of each other four and a half years ago," Joyce says. "When Danny and I got engaged I was working in the shoeshops. My father was working at Western. So was my cousin. She told me I could make thirty-five or forty dollars a week more at Western. I liked my job in the shoeshops but I quit and went to work at Western for the money."

"Me too," Danny says. "The place where I was working was doing Vietnam work and it was slowing down, so I went over to Western because the pay was good."

"We all started on the first shift," Joyce says, "but they surplused Danny to the second. I was pregnant then and I didn't like being home at night alone, so I went on second for the last couple of months. Then I took off, and stayed out for a year and a month before going back.

"I wasn't going to go back to Western because the people there were so hard to get along with. You think you know them, and then they can turn on you. I didn't want the pressure, and there was more pressure than I was used to at the shoeshops.

"They were still paying thirty or forty dollars more a week than any other place I went to, so I figured if I was going to be making money, and we wanted to buy a house, well, I had no choice. It's better this time than the last time for me.

"The area where I work now is nice. They got air conditioning, and the rates aren't that hard to make. Also I'm more relaxed. I think my yoga has helped. I used to have a hard time relaxing at work. Since I went to yoga and learned how to breathe correctly, it's been easier for me."

"I couldn't say I really like it," Danny says. "Like Joyce said, people in there are hard to figure out. Right now I'm a 34 grade, but I don't have a bonus."

"That's about equal to a 32 grade's salary," Pat adds.

"I don't see why we don't get a bonus. I'm working out in receiving. I work harder out there than I did in the plant and still we don't get a bonus. I don't really like it that much out in receiving. I had a good job in the bay area, but I got surplused to second shift, and the job I was on was too monotonous. I stayed on it a year and I still didn't get called back. So as long as I was on the second shift I asked if they could get me in receiving. It was a better job. Then I just got tired of the second shift so I asked for the first shift receiving.

"The jobs are completely different. At nights you don't do as much as you do during the day. On days you work on the trucks. There's a lot to do. I don't mind the work but they should pay us for all the work we do.

"People put a lot of pressure on you. Like this kid, he hasn't talked to me for four days because he insisted I punch his card in when I come in in the morning. I wouldn't mind so much but sometimes he doesn't show up."

"Danny could get fired on the spot for doing that," Joyce says.

"I don't know, what can I say, I don't like it. I don't think I can get ahead."

"Danny's real good at commercial art," Joyce says.

"I tried to find out about doing commercial art for Western, but they send all their work out," Danny says with a shrug.

"Me," Irene says, "I'm happy. I like Western. I like people. There's friction and all that, a lot of jealousy, but as long as I do my job, I don't mind. Right now with the new house, I have to work to help out."

"This is one of the reasons I want to get out," Pat says, "because selling I could make more than enough to keep up this house. I want more than I'm getting. I'm not getting enough, I want a better life.

"You get the feeling that there's more to life than just working at Wesern for the rest of your life. I went to a convention in Chicago earlier this year and it was really fantastic. It was something I had never seen before. Maybe in a movie, but this was real life. My first time I flew and I told myself it's not going to be the last time. As long as I work for Western I'm not going to fly too often.

"Look, I'm doing more work now than I was when I started with the company eighteen and a half years ago. My job now is more like what you'd call a material handler than a layout operator.

"I'd like to go into selling. If I started selling, I'd love it. I enjoy talking to people. At Western the most enjoyment I get is when I get the women laughing. I got to make an ass out of myself sometimes, but they laugh and that's terrific. I got several older women and a couple of them are this wide [stretches his arms]. I gotta do positive thinking on them, give them a compliment, make them think they're glamour girls.

"If I could work out the right arrangement I'd leave right now. One fellow wants me to sell cars. Another fellow started a business and wants me to go in with him, but I have to check it out a little closer.

"I told a lot of people that I might leave and they're still waiting for me to go. Some think I won't leave because I got almost twenty years of service [eighteen and a half]. The pension I'd get for twenty years isn't worth staying around, not the way I'm feeling. Other people think, 'Hey, if he goes he's coming back,' but when I go I'm not coming back. Why should I leave what I got and then go back for less than what I was making when I left? It doesn't make sense. When I make my move, I'll be gone."

A month or so passed before I talked to them again. During that time I heard from Irene that Pat had actually quit work at Western and had begun selling cars.

When I asked about setting up another interview, Irene mentioned that Pat was working long hours on his new job and didn't get home until nine in the evening. Pat had dominated the first conversation, so I suggested I get together with the other three at seven and Pat could join us when he returned from work.

When I arrived at Irene's, only she and Joyce were there. We talked about some things we hadn't really gotten to the last time.

"I stopped working when I got pregnant with Karen. She's the fourth," Irene says. "I stayed out when she came, and the others came, bing, bang, bang.

"When the older kids were younger I worked the third and then the second shift for a while. That way there was always one of us here, so it didn't affect the kids."

"My mother was never away during the day that I can remember when we were young," Joyce says. "My father and mother worked different shifts when we were real young."

"I like working different shifts," Irene says. "I mean, I've been through it so I can judge it. There's less time to argue over things between husband and wife when you're working different shifts. I think you appreciate each other more, because you see each other less. When you're on the same shift, you're both tired so you're often cross. Really we got along better when we worked different shifts."

"I think," Joyce says, "after I have my second child, if I were to go back to work, and I probably will, I'd go to work on the second shift. I think I'll go back to work after having another baby. It's just something I have in my mind, that I should work. I feel if I'm staying home the only thing I'd be accomplishing is bringing the kids up and doing housework. You have to get away from kids no matter how much you love them."

"I think Joyce is right," Irene says. "I felt it a lot when I stayed home those nine years. I wanted to get out. I felt like I actually wasn't doing anything with my life. I had to meet people and get away. I think going out and working makes you appreciate your children more when you come home. Also, when you're working you have more to talk about with your husband than if you stayed home all day long. I think your life is lighter."

"Yes," Joyce says. "You're glad to get home, you're more appreciative of your home. It seems like the people you know and your family mean a lot more to you when you meet other people.

"I've thought a lot about it," Joyce continues. "I know I'll miss Danny, but I've also thought of all these things I could be doing in the morning if we worked different shifts. Without this other person to refer to or get their opinion, you're forced to be more independent.

You're your own person, whatever you do. Growing up in a big family, I'm more a follower than a leader, and I think it would force me to make up my own mind more.

"You have to do things. That's why I liked it when Danny was taking courses, but he had to stop when we bought the house. We wanted to get a two-family house, but the banks wouldn't give us a mortgage on one so we bought a three-family. He's been working on it constantly since we bought it and I don't think that's good. That's why I wanted him to come tonight but he didn't want to talk anymore about work.

"He doesn't even talk to me that much about work, but I know a lot of things aggravate him. I think he's not as unhappy at work now as he has been because he's learning to do plumbing and stuff at home. It gives you something to concentrate on at work. I think he would still prefer doing something else."

"I'll tell you," Irene says, "I'm a little concerned economically about Pat's changing his work, but he's much happier. He's much different at home now, much more relaxed because he's doing something he's really wanted to do. He was unhappy at Western for a long time. He used to come home and it was just like 'Don't bother me.' Now he comes home and he's cheery. He's not really ready to do things because he works until nine every night, but he's more relaxed."

"I wouldn't stop Danny if he wanted to look somewhere else for a job," Joyce says, "but I know he's reluctant to change, afraid that he'd bring home less money. I don't feel that way. I know I can always get out. I know if I wanted to leave and go someplace else, he wouldn't hold that against me."

"I feel that way too," Irene says. "I don't feel the family is dependent on me, and I feel like I could quit work more easily than Pat. I'm helping out but he's still the primary breadwinner, and he feels that responsibility.

"One of the reasons I've worked," Irene continues, "is so that we could have a little more. I think the children felt that if I worked they could get a little more, it made life easier, actually."

"I don't think I ever felt that way," Joyce says. "I never had that feeling that my mother and father working together meant extra things. I don't think we were brought up to worry about material things. We lived with the idea that if we didn't have it that was it. I think maybe my parents wanted it for us more than we did."

"You feel bad," Irene says, "you can't buy them what the other kids have–the bikes and all that. They had one bike and they had to share it."

"That is one memory I cherish," Joyce says. "An aunt gave us a red bicycle with big fat tires and a red wagon attached to it. Nobody had one like that, and then somebody stole it. I remember that bike more than the bikes we bought."

"I always wished the kids could have their separate rooms," Irene says, "but we couldn't afford it. We lived in the projects for fourteen years and the kids had to share rooms. The seven girls had one dorm-like room to share."

"I didn't mind that," Joyce says. "I have happy memories of the dorm. We grew up feeling that you're never going to be alone, like you're never going to be afraid. I didn't mind living in the projects. We still have friends from those days. I remember all the good times we had even though we didn't have much money. My parents didn't own a car, but we still got to do things and it always seemed special. I really have good memories. I remember the house we moved into from the projects. It was an old house and there were bugs in the mattress that a neighbor gave us. Remember, Mom?"

"I remember. Pat found some bugs in one of the mattresses. He took a couple down to the drugstore in a jar and the guy told him they were bedbugs. We had to have the house fumigated. When Joyce found out what they were she wouldn't go to bed. She put stockings way up to here [points to the top of her thighs] so they couldn't touch her. She wore dungarees, pajamas, a hat, and mittens. She wouldn't go to bed. That was the funniest sight."

Joyce smiles and says, "Remember when you threw out the mattresses? We put them in the garbage one night because we were going to take them to the dump. A lady that we knew from the projects had her kids steal them that night."

They both laugh at the memory.

"Although I've got good memories from my childhood," Joyce says, "I want to have a smaller family. I think from being part of such a big family I don't really have my own sense of knowing what I want to do. We were never alone. We've never had to think out problems alone. What were we going to do this afternoon? You really never had to depend on yourself to occupy your own mind."

"Sometimes," Irene says, "I think maybe we had too many children, but if anything happened to any of them I'd go crazy. I came from a family of nine and I was brought up with the idea that you got married to bring up children. I mean, I'm Catholic and you figure while you're married you have children. I guess that's why I had so many. You feel as though it's your duty; you're supposed to bring up children. I think I might think differently today, maybe like Joyce says have a smaller family."

21. AFFIRMATIVE ACTION

Discrimination in employment because of race, color, religion, sex, or national origin was theoretically outlawed with the enactment of Title VII of the 1964 Civil Rights Bill. Ten years and more than a half-dozen pieces of federal legislation later, the government is still trying to make the goals of Title VII a reality. Today companies like Western Electric must do more than just show that they do not discriminate; they must submit affirmative action programs to the Equal Employment Opportunity Commission.

Western Electric, as stated at all levels of the corporation, is committed to a policy of affirmative action. The company's current president stated in January 1974, ". . . Based on my belief in WE's ability to meet great challenges successfully, I anticipate confidently the day when our equal opportunity goals are fully realized."

At the same time, on a more local level, the general manager of the Merrimack Valley Works said, "Our objective is . . . to insure that the work force profile will clearly reflect a meaningful distribution and utilization of minority group employees and to provide the full realization of equal opportunity through continuous affirmative action programs."

It didn't take me long to learn that, although the company is committed to affirmative action, there is not a very affirmative reaction to the company's program on the shop floor. In fact, there is, on the shop floor, a great deal of passive hostility toward the company's affirmative action policies.

We may be moving to a time of greater racial understanding, but there remains in the shops a great reservoir of racial hostility. Most white people I met at work didn't refer to black people as such. At best they were Negroes, or "those coloreds." An incredible number of racial stereotypes and jokes about blacks, cropped up in the most unexpected circumstances.

For example, in the midst of what started out as a political discussion about George McGovern, Richard Nixon and Watergate, a defender of Nixon's politics was pointing out that Nixon lost only one state, Massachusetts. Another person added that he had lost Washington,

D.C., as well. The Nixon supporter started to laugh and said, "What do you expect, who lives in Washington—colored people. All those coloreds are on welfare, no wonder they voted against Nixon." Nixon and Watergate were abandoned as a woman responded to this remark, saying, "You know that's right. We were down there in our car and a bunch of those colored people started throwing rocks at us."

Spanish-surnamed people didn't receive much better treatment in most of the conversations I heard. There are approximately a dozen different Spanish-speaking nationalities represented in the work force at Western, but on the shop floor they are almost always referred to as those Spanish, Puerto Ricans, or spics, regardless of their nationality.

Just as with blacks, Spanish-speaking people as a topic were likely to come up from nowhere. One evening, while I was working upstairs, we were having a smoke during a break, when an attractive young Spanish woman walked by. As she passed, Pete,[1] the fellow sitting next to me, said, "Boy, I'd like to get in her pants, but if my father ever found out he'd kill me."

One morning we were downstairs talking about baseball and someone mentioned how lazy the Puerto Rican players were. That was all that was needed to start a full-blown discussion of the negative characteristics of Spanish-speaking people. The conversation ended with one of the fellows saying, "These Spanish people, they don't want to get ahead. They like their sex; they like to have fun; they like to have a good time, and that's all." The speaker boasted that he knew what he was talking about because he had worked with them.

The language on the shop floor was often harsh and uncomplimentary. Language, however, can be deceptive. People in the shops may not employ the more socially accepted phrases like black and Spanish-speaking when referring to minority group members. This does not mean, however, that they necessarily harbor more negative feelings toward minority group members than those people who use more socially acceptable language. Racial antagonism runs deep in this society. Language, more than attitudes, separates different social classes on the issue of racial attitudes. For many of the people I knew at work the word "colored" was a descriptive term with which they had grown up, and which they often unfortunately persisted in using regardless of their feelings.

There are some people who, I'm sure, do not harbor deep racial prejudices. Probably the majority of people in the shops, although

[1] The names in this chapter have been changed.

they may not particularly like blacks or Spanish-speaking people, are not mean, vicious, or unfriendly toward specific minority group members. People have a way of differentiating between groups and individuals. I don't know how many times I heard someone say something like, "Oh sure, I like Gladys, everyone does. She's not like the rest of those coloreds."

Similarly, a person I knew who had nothing good to say about Spanish-speaking people was very helpful to a Cuban woman who worked near us and was taking company-sponsored English classes. He seemed genuinely to enjoy the experience of helping her practice her English.

Amid all the general racial antagonism that existed, what people seemed to resent most bitterly was what they considered the company's bending over backward for minorities.

Western Electric remains, according to most people who work there, one of the best places to work in the area. Therefore, many workers are anxious to get their children, relatives, or friends into the plant. A number of people told me that if you aren't a "minority" it is becoming increasingly difficult to get a job in the shops.

One night I was eating dinner with a couple of fellows when some Spanish-speaking men sat down at a table near us. I didn't think anything of it until they left. Bill Mellen, an older fellow who was sitting next to me, was very upset. He said, "I understand the company is far behind in their minority hiring and that the colored and the Spanish deserve a chance to work here, but they're going too far. The company pays for the Spanish to go to English classes. If they have any problems or need help, they give it to them. They give them preference in hiring. Now my kid wants a job in here, and he can't get one because they give preference to minorities. How about the company's obligation to me? Don't they have some obligation to me? I've put in fourteen years here. I haven't asked for much. Now I want my kid to get a job, and he can't because they're hiring minorities. Sure it pisses me off. It would piss anyone off."

People believe that not only are minority members given preference in hiring, but they enjoy special treatment when it comes to promotions. A woman told me, "You know, when I started working here they used to have a test before you could become a tester. That was before they started hiring minorities. When they found out the minorities couldn't pass the test they dropped the test so that they could promote them."

Shortly before I left work several people were upgraded. When the selections were made known, there were several people who were especially upset. One told me, "It's getting crazy around here. They're promoting coloreds, Spanish, women, Lesbians, and switch hitters. It's this equal opportunity act, you gotta hit both ways if you want to make it anymore around here."

Ted Kelly told me, "They've got a new slavery in here. It's the white man who's the slave now. In here if you're white, and especially if you're a man, you might as well forget about getting ahead."

One of the people who received a promotion at that time was a Cuban woman. She came over to our department, after the announcement, to talk with a few of the women who used to work with her. After she left, Alice Greggins told me, "You know, Rosa is all right. When I was in her group I didn't like her at all. I didn't like a lot of those Puerto Rican women. They don't have attitudes like ours. This woman only came here from Cuba four years ago and she's just got a big promotion. Now there are a lot of people like me who have been here much longer, and we haven't got the promotions. The company just bends over backwards for these people and it isn't right. These Puerto Ricans they come in with an attitude. They know the company will bend over for them. They'll tell a supervisor that they don't want to do certain jobs, and they want other work. If we came in with an attitude like that they'd just get rid of us, but not these minorities. They get away with murder. The company bends over backwards for these people and it isn't right. They don't bend over for us.

"Some of them are OK, once you get to know them. Like this woman. She had a tough time, back in Cuba. Castro took all her property, I guess she was pretty wealthy. They took a raft over to Florida and then came up here. She's got a college education. I like her, like I said, and we get along. But a lot of them have strange attitudes and I don't think it's right."

The company has set up classes in English as a second language in the plant. The classes are offered to employees whose native language is not English. The object of the course, according to a company bulletin, is for people ". . . to be able to understand, speak, read, and write sufficient English to perform in entry level jobs, safely and proficiently." Enrolled workers are given two hours a week off from work to go to classes.

When I first heard about the course I thought it was a good idea and one that few people would get upset about. However, to my surprise a large number of people had something snide to say about the classes. One older Italian fellow said, "Can you believe it, they give those Puerto Ricans paid time off to learn English. When my father came here, he didn't speak any English. You think anyone paid him to learn it? Hell, no. Nobody did a goddamn thing for him. Why should they get this special treatment?"

This program more than any other run by the company was singled out as an example of how the company doesn't do anything for "us" but does all kinds of things for the Spanish.

The last rumor I heard about minority employees receiving special treatment was that the company was trying to protect minorities against layoffs. I was told to look at Article 28 of the new union contract. At the union's contract ratification meeting a shop steward from the second shift got up and said some of his people had asked him if Article 28 was inserted to protect minorities. The local's president assured the person that the exception had not been inserted to protect minorities, but to handle what the company considered certain business emergencies. [The contract states, "The company may, however, exempt from selection for layoff certain employees when such exemptions are necessary to avoid unreasonable departmental depletion."]

With everything that had been said to me about minority employees, I was surprised that of the approximately 5,700 shopworkers between grades 32–38, only about 265 are minority group members.[2] The minority workers are predominantly found in the lowest work grade positions. There are approximately 188 minority workers comprising slightly less than 10 per cent of 2200 32 grade workers. The percentage is less than 4 per cent at the 33 grade level and less than 1 per cent at the 35–37 grade levels.

The minority representation at higher levels is even smaller. Of the approximately 575 officials and managers, four—three section chiefs and one department chief, less than .7 per cent—are from minorities.

The Spanish-surnamed Americans are, more than any other minority group, disproportionately underrepresented in higher positions. Although 180 of the 265 minority employees at grades 32 through 38 are Spanish surnamed, only nine of them are in positions above 33 grade. Furthermore, not one of the four minority employees in supervisory positions is a Spanish-surnamed American.

[2] Blacks, Spanish-surnamed people, Orientals, and American Indians are taken to represent the minority population.

The relatively few minority employees in the shops reflect the area's general population. Unlike many of America's large cities, the Merrimack Valley remains an area with a relatively small minority population. There are 230,000 people in the immediate statistical region, and of those the minority population is 13,000 or approximately 5.5 per cent. Nearly 85 per cent of that minority population (11,000) is Spanish surnamed. Fifteen years ago the Spanish-surnamed population in the area was approximately 1500 people.

Seeing the statistics made me wonder about a number of things I had heard in the shops. Were minorities receiving preferential treatment? And why, if there are relatively so few minority workers, do people seem so resentful of the company's programs?

First, while the absolute number of minority employees is still fairly minute, the number of minority workers representing new hires is much greater. So even though minority employees may represent less than 5 per cent of the graded work force, they may represent nearly 20 per cent of new hires.

I thought perhaps the same might be true for promotions. Even though minority employees might make up a negligible part of the existing supervisory personnel and higher graded shop positions, in recent promotions they might comprise a significant percentage. More minority employees are being promoted. However, for the immediate future their numbers will remain fairly small. For example, there were approximately 450 36 grade workers in 1973 and only six were minority group members. For 1974 the company projected the need for 136 new 36 grade workers, and of those only three will be minority employees. A quick look over at supervisory positions, where promotions are not as clearly tied to seniority, shows that there will be some increase in minority group representation, but still not in large numbers. Therefore, even though the company has an affirmative action program, the number of people immediately affected is relatively small.

Even if the statistics didn't bear people out in regard to promotions or new jobs, I thought that maybe in other programs there was favoritism. Since I had heard so many complaints about the English as a Second Language (ESL) course, I took a look at the past year's enrollment, expecting to find that more than 90 per cent of the participants would be Spanish speaking, because I had never heard anyone complain about any ethnic group taking the ESL courses but Spanish-speaking people. The figures showed 60 per cent were Spanish speaking, 17 per cent Greek, 10 per cent Italian, 7 per cent Oriental, 2 per cent French-Canadian, 1 per cent Polish, and 3 per cent other Europeans.

A union official told me he had repeatedly heard remarks from people on the shop floor about how blacks and Puerto Ricans with almost no service were given instant promotions, thereby skipping the usual qualification-service-connected promotion scheme. He kept on telling people that such promotions were impossible and violated the labor contract, but the rumors persisted. Finally he said he got so disgusted with hearing what he knew were unfounded rumors that he walked into the shops with fifty dollars. He went around challenging people to show him a worker who had gotten such preferential treatment. He said he'd wager his fifty dollars against five dollars that they couldn't show him any instant promotions. He said that not one worker took him up on the bet. However, he added, "I knew when I walked out of those shops that some of them still believed there were people getting instant promotions."

A company official who was quite concerned about the persistent rumors of favoritism, preferential treatment, and bending of rules to aid minorities told me that practically all of what I had heard was unfounded. He said no tests had been dropped or rules changed to accommodate minority workers. Furthermore, he claimed that it wasn't true that unqualified minority workers or women were being promoted. However, he admitted that if two people were equally qualified the minority worker might be given preference. Despite all the assurances that company and union officials give, large numbers of people still believe minority group workers are receiving all sorts of preferential treatment.

These prejudices are deep-seated and historically traceable to the area's ethnic cleavages. The Merrimack Valley in the late 1800's and early 1900's was the recipient of several waves of ethnic migration. Large numbers of Irish, Italians, Poles, Eastern European Jews, and French-Canadians came to the area to work in the textile mills and the shoeshops. These were distinct ethnic communities with deep ethnic antagonisms. Old newspapers are full of reports of how ethnic antagonisms flared into violent disruptions.

Many of the old-timers in the plant talk about the old days and the strong ethnic churches and gathering places like the Irish Hibernian Halls. Today the ethnic divisions and antagonisms are much less strong. They have been replaced by racial antagonism.

Although racial prejudices should not be dismissed, they are not the only reason for worker opposition to affirmative action. One cannot overlook the fact that many people feel that they will be asked to pay a real and immediate price for the advancement of minority group members. There are not an infinite number of opportunities open at Western Electric or anywhere else for that matter. Many of the people in the shops believe that these opportunities will be further diminished by the company's affirmative action programs.

This society has discriminated against racial minorities, but it is also true that it has not been particularly generous toward ethnic Americans. Great numbers of people working in the shops are first-, second-, or third-generation immigrants. Many of the people I met at work have told me stories about parents or grandparents who came to the area's textile mills and shoeshops speaking no English. They were paid low wages and given no assistance as they struggled against an often hostile environment. It is not surprising then that people whose parents or grandparents suffered through an unsupportive environment would be bitter toward people who are receiving what they perceive as advantages that their relatives did not enjoy, and that they themselves are not being given.

Affirmative action often places minority workers in an uncomfortable position. I wondered how they felt about the situation, but I met very few minority workers while I was at Western. There were two minority employees in my group on the evening shift. There were none where I worked on days. Now, a year later, there are four.

I avoided asking the few blacks or Spanish-speaking people I met specifically about their feelings on the racial situation in the shops, just as I rarely approached whites about these matters. What I have reported here are things I heard people talking about spontaneously.

So I have very little information about "the minority population's feelings." One Cuban woman told me she didn't speak English so well, so she spent most of her time with other Spanish-speaking people. The most substantial information about how minority workers felt comes from a couple of people I got to know after I stopped working. One woman was quite frank about her own feelings and about what things were like. She felt there were a lot of prejudiced and bigoted people at Western Electric. She thought the company was putting black and Spanish-speaking people in a nearly impossible situation. She said, "Sure there's racism and hatred in here, and the company has added to the hostility. If they had done what they should have done in the first place rather than just running around saying they were an equal opportunity employer, you wouldn't have nearly the problem you've got today. They didn't do anything until the Equal Employment Commission came in and told them to get off the pot, to get some minorities in here. So now they're bringing in the minorities, cramming them down people's throats. That's why there's so much hatred in here. It made an already bad situation worse."

As individuals, we may not be sure how we want to handle "the minority problem," but the courts and Congress have decreed that in the area of employment companies shall no longer have a free hand in deciding how they shall proceed. A company like Western Electric can no longer meet the requirements of the law by simply stating that it is an equal opportunity employer. Companies are now being required through a program of affirmative action to compensate for a history of either intentional or unintentional discrimination. Quotas for hiring and promotions at all levels must be set and met. Yet, this necessary program has some negative aspects, such as the pitting of workers against each other. In many white workers' minds minority employees are villains in league with the company and the government to take away the few advantages that whites have.

If the majority of the workers are hostile and bitter about the company's affirmative action program, it is regrettable but understandable. Their anger cannot be dismissed fairly as simple racist feelings. Much of their anger stems from their anticipation that they will be made to suffer for social change. Many people believe that in employment, as elsewhere, they are asked to pay disproportionately for social change. Wealthy people in this society are often isolated at home, in schools, and at work from "the minority problem." Working-class people are not.

22. JUNIOR VALENTIN

JUNIOR began to work in my department after I had stopped. Over the months I kept coming in, we started talking. I wanted to document the experiences of a Spanish-speaking worker but didn't want to have management find me a "typical one." Meeting Junior provided an opportunity to look at one person's experience and views.

Junior Valentin moved out of his family's home three years ago, when he was nineteen. He lived for nearly two years in a one-bedroom apartment in Lawrence before he and his girl friend, Michelle, moved into a more expensive, modern, one-bedroom apartment. After a year of splitting the $170 monthly rent, Junior decided it was time to move again. Michelle moved back in with her parents, and Junior moved over to his current apartment. He pays twenty dollars a week for a room and shares the use of the apartment with his friends, Diane and Roland Rivers.

Junior relaxes in the apartment's living room and talks about first coming here. "My father came here ten years ago because he lost his job. He worked for a petroleum company in Puerto Rico and they laid him off. He didn't want to collect unemployment, so he came up here to my uncle's house and he started working with my uncle at Converse Rubber.

"Two years later the whole family moved up here to be with him. I didn't really want to leave Puerto Rico. I was only fourteen and all my friends were there, but I moved because I always liked to be with my family.

"Even now I think of them all the time. They moved back to Puerto Rico, but I stayed up here. My father always planned to go back. We have a big house and a farm there. He only came here to make money for the house, and six months ago when he had enough they went home. They are waiting for me, but I'm not really sure if I'm going to go back to Puerto Rico.

"My father didn't ask me if I wanted to go back with them when they left. He knows I'm old enough to know what I want, and if I wanted to go with them I would have said so. I respect my father and

I love my family, but when I want to do things myself they know I'm mature enough. In Puerto Rico when you are eighteen you're old enough to make your own life. Here you have to be twenty-one.

"I've always liked to be independent, doing things for myself. I've always liked to have money in my pocket, honest working money. I had my first job when I was twelve. I worked in a supermarket. After that I worked in a restaurant and then a gas station. My father bought me things when I wasn't working. When I started working, though, I told him I'd rather buy what I need on my own.

"Some kids grow up with their father giving them anything they want. They make their children real spoiled, and that's no good. You have to make your sons so if they want something really bad, they do something for it. They do not rob, but they work hard. That's the way I was brought up. That's how a lot of Spanish kids are raised. Some Spanish kids grow up just like you white people, spoiled. I don't really talk about races, though. It all depends on the family you are with. If the family is a spoiled family they are going to spoil you.

"I went to work when we first came here. Going to work was my own decision, not my parents'. I wasn't even sixteen so I had to get working papers from the school. I was making sixty, seventy, eighty dollars a week depending on whether I was working overtime or not. That was when wages were only $1.70 an hour (1967).

"I think I worked for a year and a half and then I went back to school. I didn't stay long. I had to quit because I needed money. My father was the only one working in the house. There are ten children in my family; so it was kinda hard for me being the oldest. I quit school and went back to work at Lawrence Maid Footwear.

"I didn't like working in the shoeshops. I wanted to get out and get a better job really bad. I tried to get into Western, but they wouldn't let me start until I was eighteen. I made out an application, and they told me to come back after my birthday.

"I started working at Western in September 1970 on the second shift so I could go to school days. I did that for a year and then I had to give up one of the two. I used to get to bed at one o'clock in the morning after working and then I'd have to be up at seven to go to school. I was only getting six hours of sleep, it isn't enough for a workingman and student at the same time. I gave up school. I only had one year left to go, but I can always finish that one year.

"After I quit school I tried to switch onto the day shift. I had worked nights for too long. I wanted my evenings free. When an opening came along I took it. Maybe two months after I switched onto days I got laid off.

"I was out for about eleven months. I didn't go to work. I collected sixty-nine dollars a week for eleven months—that's almost what I made in the shoeshops. Then they called me back. I went back to work on the second shift assembling channel banks. I worked there for about a year, then something opened up on days and I moved.

"This is a good group, but I like the work I did on the bays more than on this job. On the bays you work on a big thing. I'm a lot better on the big things. Here you have to work with small things all the time and I don't like it.

"I'd like to get off the bench and get a really good job. Some of those jobs you need an education. I have education but not good education. For me a good job would be a technician or tester, but I don't have enough service. I think I have enough service to be a material handler, and that's a good job. I'd like that because I wouldn't have to be sitting down like I am now. Sitting down all day isn't my type of work.

"Right now I'll probably stay where I am, and take care of the bills I have. I used to save some money to buy U. S. Savings Bonds. I think they took out $12.50 a week from my check for savings bonds. Here, I'll show you."

Junior momentarily disappears and reappears with a handful of twenty-five-dollar U. S. Savings Bond.

"I don't know, I guess I got about thirty of them. I won't touch them for a long time, but now I can't save anymore for bonds, because I have a lot of bills.

"I just bought a car (a green MG) through the credit union. They take twenty-five dollars a week out of my pay for it. They also take another seven dollars a week for a loan I had before this. A couple of years ago I borrowed three hundred fifty dollars. I used a hundred to buy a 1965 Catalina. I used another hundred to fix it up, and the other hundred fifty for some clothes.

"I'm also paying for a twenty-three-inch Admiral color TV. I'm paying back whatever I feel each week. Sometimes I give him [the store owner] fifteen dollars, sometimes twenty dollars. I want to get rid of it. I only owe him seventy-eight dollars. I don't know how much the interest is, it's somewhere on my payment book. I think it's about fifty dollars.

"Right now I want to pay off my debts. I don't like to have a lot of

bills over my head. You can go crazy with a lot of bills after a while. When I get them paid off I'll start saving again. I figure if I go to Puerto Rico I want to go into a business of my own. I don't want to work for nobody. I'd like to own a boutique or something like that. I like nice clothes and that way I could sell them and save on them too.

"I like to dress nice. I'm a neat person. If I worked in an office or something, I'd always wear nice clothes. But you don't have the job, you don't make enough money to wear nice clothes all the time. You have to do with whatever you have. Even now I like to buy nice clothes. The most expensive things I get are shoes. The most I've spent on a pair of shoes is fifty-five dollars. You can spend more, but I don't like to. I have about five pairs of shoes. I also try to get nice shirts and pants. I like to look nice.

"I don't like to spend too much money, but if I see something I like and I can afford it, I buy it. I feel I should treat myself to nice things. I'm just trying to live. Even since I had a bad accident a couple of years ago I said if I live this is going to be a second life. I missed an exit and went straight through the guard rail. The car, my 1965 Catalina, rolled five times. I was still conscious. The car was totaled but I got out of the car and walked away from it. I figured then after that I'm going to enjoy life, take it easy.

"I'm young and I've got plenty of time to make my life what I want it to be, so I don't worry about whether I can get what I want today. If I can't get it today, I'll get it tomorrow, or the day after. One of these days I'll get my chance to make it. If I don't, then I don't. I don't worry about it.

"I'm happy the way I am. As long as I have my job I'm all right. I do OK. I'm not going to be wealthy, but I don't want to be wealthy. Wealthy people don't have things or thoughts I have. For me, I just want enough to live on and be happy.

"I'd like to get married, have children and raise a family, but right now I don't want to get married. My girl friend, she might want to, but not me. I'm too young to die." [He laughs.] I'm very independent. like to do things my own way. Like if I want to go out, I don't have to tell somebody I'm going to go this place or that place. If I'm going to go out, I just get dressed and go out, and when I come home, I come home. I don't think I want to get married until I'm twenty-six or twenty-seven. It's tough when you get married young, and I don't want to screw up my marriage. If you get married or screw it up you'll probably be married many times before you're dead. I don't want to be like that. When you're ready to settle down, you're ready. When you're not, you're not, and I don't feel I'm ready yet.

"When I get married I'd like to have a nice house, a piece of land and children. I'm Catholic and the church says you're not supposed to avoid having kids. I don't feel that way at all. I think in this life we're living, it's very expensive. Forty years ago it was very cheap, but now I don't think the young generation can afford to have so many children. I think as much as they can have is four, unless they are very wealthy. For me, I don't think I'll have more than two. For me I think it's very tough so I want a small family.

"It takes more money now to have a family, but I'd never want to work two jobs to do that. That is killing yourself slowly. You can't enjoy what you're working for. Let's put it this way, you take a horse and you put three hundred pounds of something on that horse every day, year after year. He's going to be dying little by little, and by the time he gets old he just says, 'I'm gone.' If a person does too much he's like that too. You work so hard, and when you're old, you wouldn't be able to enjoy it. Working like that is working like an animal. You should work, there is no choice, but you shouldn't kill yourself."

23. WOMEN AT WORK

On one of the first visits to the Merrimack Valley plant I was given a tour by someone from the public relations department. As we passed one department my guide said, looking out at row upon row of women doing benchwork, "This is a good place to work, especially for women. I wouldn't mind my wife working out in the shops."

While working in the shops I heard a variety of comments about women. "Women don't want to be promoted; women are better equipped to do benchwork; women aren't as bored as men by benchwork; 32 grade jobs pay a decent wage for a woman but not for a man; and, nowadays, if you want to get ahead at Western you've got to be a minority or a woman."

I wondered what all these statements meant, and why women came to work in the shops. It didn't take me long to learn that, unlike the highly publicized group of middle- and upper-middle-class women

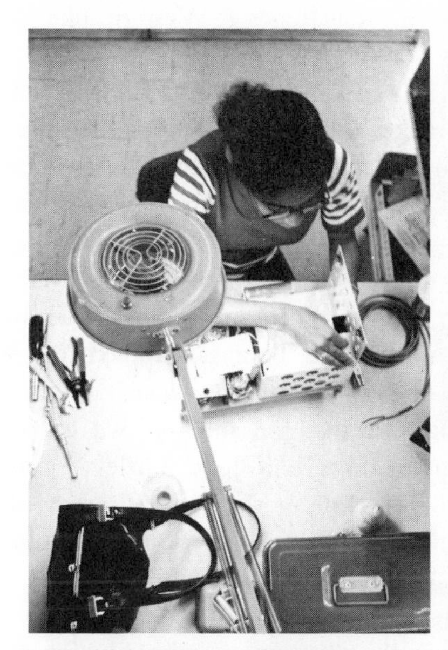
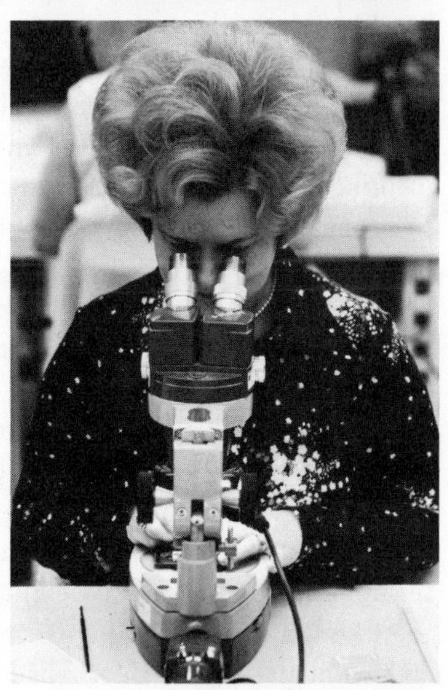
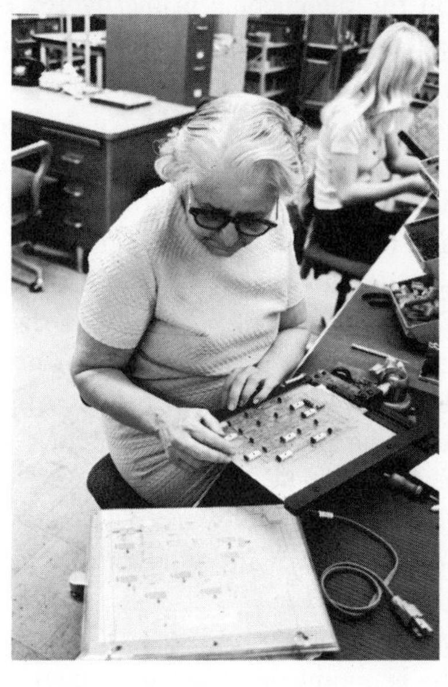

who have been going to work in increasing numbers to find careers and a sense of personal fulfillment, as well as to earn money, the majority of women who come to work on the benches at Western, like their male counterparts, do so primarily for economic reasons.

"I could have worked upstairs in the office," Claire[1] told me, "but they don't pay what you can make out on the floor, especially if you're in a good department with a good bonus. Maybe the work isn't as interesting, but it isn't bad and the money is decent, especially for a woman."

Some of the women I met at work are primary breadwinners. However, the majority were working, they said, to supplement their husbands' incomes. For many women, like Mae Thomas, supplementing a husband's income means making enough to get the necessities of life. "I came to work," Mae said, "because my Joe and me, we couldn't make it on his salary." For many others like Jean Turley, supplementing means making enough to get some of the material extras that this society offers. "Sure we could get by," Jean said, "but we couldn't have the two cars or the boat. When we pull that boat out to the beach I'm really proud that I've helped make it possible."

Not all the women who work in the plant come primarily for economic reasons. Some come to get out of their homes after their children are grown. One woman, Lois LeFeber, told me, "I had to get out of the house. I was fat as a blimp. My kids were grown, and all I was doing was running between the kitchen and the TV. Watching soap operas and eating was all I did. That's why I came to work, and I feel much better. I've lost weight and made a lot of friends in here."

Although sex roles are undergoing certain changes, most of the men and women I met at work had fairly traditional views about the responsibilities of men and women. The man was, if possible, to be the primary economic provider, while the woman's principal responsibility, even if she worked, was the home and the children.

Several men I met at work felt the need to be the sole economic provider. Terry Thomas told me, "Both my parents worked when I was young and it hurt me. I work two jobs so that my wife can be in the house with the kids where she belongs."

Generally, men who work two jobs are applauded as being ambitious and willing to sacrifice for the economic well-being of their families. A man who moonlights by working sixty to seventy hours a week is usually not doing it to find the job satisfaction in a second job which

[1] The names in this chapter have been changed.

might be lacking in the first job, but rather to be a better provider by bringing in more money. A man who does this dramatically limits the time he spends with his wife and children. On the other hand, women who work even one job are frequently criticized for neglecting their husbands and children.

Women who work in the plant, like most working women, have a second job of wife, mother, and housekeeper when the paid working day is finished. These women cannot generally afford a helper to come in and do cleaning or cooking for them. So after each day's work there was dinner to prepare and cleaning to be done. These household responsibilities leave little time for relaxation and "self-development."

Jane Donahue, a fifty-year-old woman, told me she would like to take an adult education course but had no time. "I work here eight hours and then I go home and prepare dinner. After dinner I do the dishes and laundry, and if there's enough time I iron some shirts for my husband or my boys. There's no cafeteria where they work so I make lunch for them.

"By the time I'm done with all that I'm exhausted. I'll either sit down and watch TV or go to sleep. Sometimes I go to sleep watching TV. I never have time even to see a whole movie. I've got to be up at five the next morning. There's really no time for anything else."

Some women are helped with their household chores. I heard women talk about their children or husbands doing cooking and cleaning. Marlene Peters told me, "When a woman works, every member of the family needs to chip in."

Vicky Calo is surprised by her husband's efforts. "I didn't believe my Mike when he said he'd do the dishes and wash the floor, but he has." Others like Ann Hope expect help from their husbands. "It's only fair," she said. "I'm helping Ralph by working, he should help around the house."

As much as children and husbands helped out, the majority of women I met at work accepted, whether they liked it or not, the traditional division of labor by sex.

JoAnn Farley mentioned that her husband had hurt his leg, and although he drove her to and from the supermarket he couldn't help her carry the bags in. She said it was the first time in the thirty-two years of their marriage that she could remember ever having to carry grocery bags into the house, and she knew how bad it made her husband feel. Janice Vitale laughed, and said, "You're lucky. When my husband is home he asks me to help with the outside work. One day I was helping him move something, and a neighbor asked if he could help. My husband, the big shot, said no, I would help him. I did, but do you think in the house he could pick up his socks, or hang up his clothes? Nope, he comes in the house, and where he takes something off that's where it stays. I walk in the house after work and I've got to clean up after him like he was a little kid. What are you going to do? I guess they're right, you can't teach an old dog new tricks."

If they were not employed, most of the women would take the overwhelming responsibility for their children. Many have, in fact, not held jobs when their children were young, believing they should be home with them. Large numbers of women have returned to work only when their children were of school age.

However, there are women with younger children who work. Once a woman with children goes to work, the question of responsibility for child care becomes immediate in a way it rarely does when a man goes to work. A woman who goes to work and has children must assure others that she is not neglecting those children, that adequate arrangements for their well-being have been made.

I was surprised to find how few factory workers I talked to availed themselves of professional day care services. Many single female parents and married couples who worked the same shift leaned heavily on parents and in-laws for child care. I met many couples who chose to work different shifts in order to take care of their children. I thought the reasons were that professional day care or baby-sitting was prohibitively expensive, that it would eat up too much of a second income or that parents and in-laws were not available.

This would imply that less expensive day care programs would make working different shifts unnecessary. I learned from several couples that, money matters aside, they were reluctant to allow professionals to raise their children, especially when they were small. "We brought them into this world," Ed Patrick said, "we figure we ought to raise them." Ann Dowd was resigned to the price she and her husband paid in working different shifts. She told me, "I don't want any strangers raising my children. Sure we don't get to spend much time together working different shifts, but we won't always be doing it this way."

Some couples who worked different shifts maximized the time they had together by having one work the first shift and the other the third, thus giving them six hours of the day together. It was, however, not uncommon to meet couples where one parent worked the first shift

(usually the man) while the other worked the second shift (often called the mother's shift), thus leaving the couples less than ten minutes a day together. Liz Walker told me, "By the time my husband comes home I'm already gone. By the time I come home he's already asleep. It's not really a very good system, is it?"

Women, I found, unlike men, spent much of their time at work sharing information about child rearing, discussing new household purchases, and trading recipes. I think the amount of time they devoted to this not only showed how important their children and homes were, but expressed a subconscious need to bring their families to work, to present themselves as adequate wives and mothers. Men infrequently talked about their families.

Most women in the shops at Western enter as 32 grade operators. This is the lowest graded, supposedly least skilled, benchwork. The starting salary for a 32 grade operator is $3.14 (1973) an hour. Three years ago it was $2.68. On top of the basic wage there is a bonus that, plantwide, averages about 25 per cent, so many of the people working at the lowest grade are averaging approximately $4.00 an hour.

Of the approximately 5,500 workers who are in the shops (grades 32–38) approximately 2,000 are 32 grade workers and of those 90 per cent are women. For a long time only women were given jobs as 32 grade employees. Men were started at 33 grade work. Although there was no company stated policy, many people said this was the procedure because 32 grade work was not considered man's work, and the wage paid, though considered good for a woman supplementing a man's income, was not considered a decent wage for a man who needed to support his family. This procedure was discontinued in 1970 and men as well as women are now given 32 grade jobs.

Out in the shop many people still hold a feeling that benchwork, especially that of 32 and 33 grade operatives, is woman's work. Men frequently commented that women were better equipped to do the benchwork. I wondered if they meant that women could perform the tasks demanded of benchworkers better than men. From watching people work for five months, the answer is a qualified yes. In our department we worked primarily on printed circuit boards with small parts. Women seemed quicker at completing these boards. I heard many explanations for this. The most common, and the one that made the most sense, was that women have smaller fingers and consequently greater finger dexterity.

Within the limited range of work assignments in our area, there was a sex separation of tasks. Given the chance, our supervisor would put the 32 grade men on bigger boards, boards with large transformers, which, he claimed, the women didn't like to work on. Both the men and the women in our group seemed to agree that bigger boards and jobs calling for greater physical exertion were men's work, while the jobs calling for work with small parts were women's work.

I worked primarily on the big boards. A couple of women told me they were glad when men started coming in the department because they used to have to do the big boards, and they hated them. "It's too tough for a woman," Jackie said. "I'm glad Jack has you guys working on them." Diane told me she used to work on the big boards and she wasn't, she said, "strong enough to do the work." Just as most women seemed to prefer the smaller boards, the men seemed to like the big boards. Larry said, "I don't think there is anybody who can do these big boards any quicker than me. I don't like the small boards," he continued, holding out his hands. "My fingers are too big, too clumsy, to work on those small boards."

Our work load didn't always permit the men to work on the large boards. We were often required to do small boards. I felt more confident on the larger boards than on the smaller ones. Like the other men, I often complained about working on the small boards. I found I was saying how big and clumsy my fingers were. What was I saying by complaining, except that it was woman's work?

I realized after working a while that, when men talked about women being better equipped to do the benchwork, they were not merely talking about finger dexterity. They were talking about a belief that women have a better capacity to tolerate the boredom and repetitive nature of much of the work.

A man who worked in the tool room told me he hated it when he used to do benchwork. He said, "It was boring as hell. It bored all the guys I knew, but for some reason it didn't seem to bother the women. I don't know, once in a while one of them will tell you they don't like it, but the majority–I'd say 99 per cent of them–don't seem to mind. I don't know if it's physical or what, but women don't mind the work and men can't stand it.

I was convinced by the end of my work stay at Western that women were much better at adapting to the boredom and repetitiveness of their tasks than the men. They might not like it any better, but they seemed to accept the boredom more readily. I asked a woman if she thought I was silly to think that men had a more difficult time adjusting to the work than men. "No," she said. "It's much more difficult for men to be submissive and passive and that's why they have a harder time adapting."

Ruth's remarks made a lot of sense to me on the surface. It was easy to believe that the attitudes of both men and women toward benchwork reflected the different socialization patterns usually associated with men and women. Men in this society have been taught to be aggressive and competitive, while women have been taught to be submissive and passive. It would seem to follow logically that women should be better equipped to tolerate long hours of work that is not intrinsically interesting.

Although the explanation made some sense, the more I thought about it, the more it bothered me. If women may be more able to adapt to boring, repetitive work, men still do some of the most boring, routine industrial work. For example, it's hard to imagine many more tedious jobs than those performed by large numbers of men on the assembly lines in the automobile factories.

If the idea of women's work cannot be explained by finger dexterity or the easier acceptance of boring work, I wondered what made certain jobs woman's and not man's work. At Western I think some of it has to do with the physical requirements of the work. Work that requires physical labor and is dirty is considered by both men and women as man's work. Work handling small parts with little or no physical exertion is woman's work.

The wage is also, I discovered, important. Although men were disturbed by the boredom and repetitiveness of the nonphysical, low-paying benchwork, they were willing to work on equally dull jobs where the work is physical and the wages are better. So even though the 32 grade work is now open to men as well as women, it is still widely thought of as woman's work.

Joanne Bellini has worked for the company for four years. She said that she and her husband started work on the same day. After a week her husband decided to quit, saying it was woman's work. She didn't disagree with him. She stayed while he went out looking for another job.

Many women, especially older ones, openly stated that they thought the lower-grade benchwork was woman's work. As much as they themselves might dislike the job, these women often expressed sympathy for young men who were, as they said, stuck doing 32 grade benchwork. Grace Mazoola told me that although she wouldn't mind her daughter coming to work on the bench, she didn't want her son to because it wasn't a job for a man.

Whatever their feelings, relatively few women have moved up the graded system and into supervisory positions. As one goes up the graded skill ladder, the number of women dramatically decreases. Although women represent nearly 90 per cent of the 32 grade workers,

they represent only slightly more than 50 per cent of the 34 grade workers, 28 per cent of the 36 grade workers, and only 20 per cent of the 37 grade workers (April 1973). Until 1970 there wasn't a single woman supervisor. That year ten women were made supervisors. Today there are approximately fifteen.

The company acknowledges that women, though unintentionally, have been precluded from advancement and are strikingly under-represented in higher paying and supervisory jobs, much like minorities. But the company considers the problems of women and minorities somewhat different. Whereas the minority population is relatively small in the immediate area, nearly 50 per cent of the local work force is women.

The company wants to rectify this situation. In an internally distributed document the company states, "All jobs are open to all of our employees and we expect that gains will continue to be made especially in the case of females, where attitudes previously held regarding jobs are changing in a positive direction. We will attempt to accelerate this change by encouraging females to consider all available job opportunities and by publishing the fact that we have no jobs that are considered 'male only.'"

The company is anxious to find qualified women and move them ahead. There are many women who wish to get ahead who feel, with some justification, that they have been held back or ignored because they are women. There are a great number of women who might be interested in promotions but who have a major concern that often makes them hesitate. It is a concern to which most men don't seem to give priority: their children. Whether a woman is married, separated, divorced, or widowed, if she has children, those children will frequently cause her to hesitate about any promotion that will require her to change a shift or will demand more of her time. Similarly, many of the older women whose children are now grown, and who have been working to help husbands build a little nest egg for retirement, are reluctant to take a promotion if it requires that they spend more time away from their husbands.

For a long time many people in the company said that the reason there weren't more women in higher positions was not so much the company's policies or omissions but that women didn't want to be promoted. A supervisor I knew, who had worked for the company for twenty-seven years, told me that in his experience women were much more reluctant than men to accept a promotion. "I don't know," he said. "Offer a man a few extra pennies and he will take a different department, a different shift. He wants to get ahead, but women for a couple of extra pennies will worry more about their family and won't move so quickly.

"It isn't a question of men as opposed to women. It really is who is the primary breadwinner, and who is the secondary breadwinner. If the woman is the primary breadwinner, she'll move, but still she'll move less quickly than most men."

I wondered if this were really true. Over the months I talked to quite a few women and men about promotions. Most of the men I knew were willing to take most promotions if it meant getting a higher grade and a little more pay even if it meant rearranging home schedules. I met a man with four children. He had worked days for twelve years and only recently had switched to nights. "They offered me a 35 grade on nights, and I thought maybe I could get to 36 grade quicker working nights. I don't like it but with the extra money the night bonus [10 per cent] and the chance for further promotion, I felt I couldn't say no."

The attitude among the women I knew was generally quite different. Several women who had long service records told me that they had quit work more than once during the years because they were switched to another shift. "I was working on the first shift," Beth said, "and they bumped me. They were going to send me to the second shift. It meant I couldn't be with my kids [then school age] so I quit. It made it tough on us, but I waited until they had an opening on days again. I was out of work for more than a year." Several women I knew told me if they were surplused or bumped onto another shift they would quit rather than take the transfer.

Even when a woman is the primary breadwinner, she often has to arrange her shift according to her family, and she feels less freedom than a man to switch shifts because of children. I knew one woman who had worked nights for nine years. She was divorced from her husband and needed to work. She took a job on the second shift, not because she wanted to, but so that she could be home during the days with her children. Once they were in school she switched to days. She would like to be promoted, but if a promotion came along that required her transferring back to the second shift, she would not accept it, because she wouldn't have time with her now teen-age children.

Many women didn't want to do "men's work." Unlike the men, the women frequently mentioned a clean environment and not being physically tired as important aspects of work. Many women came to

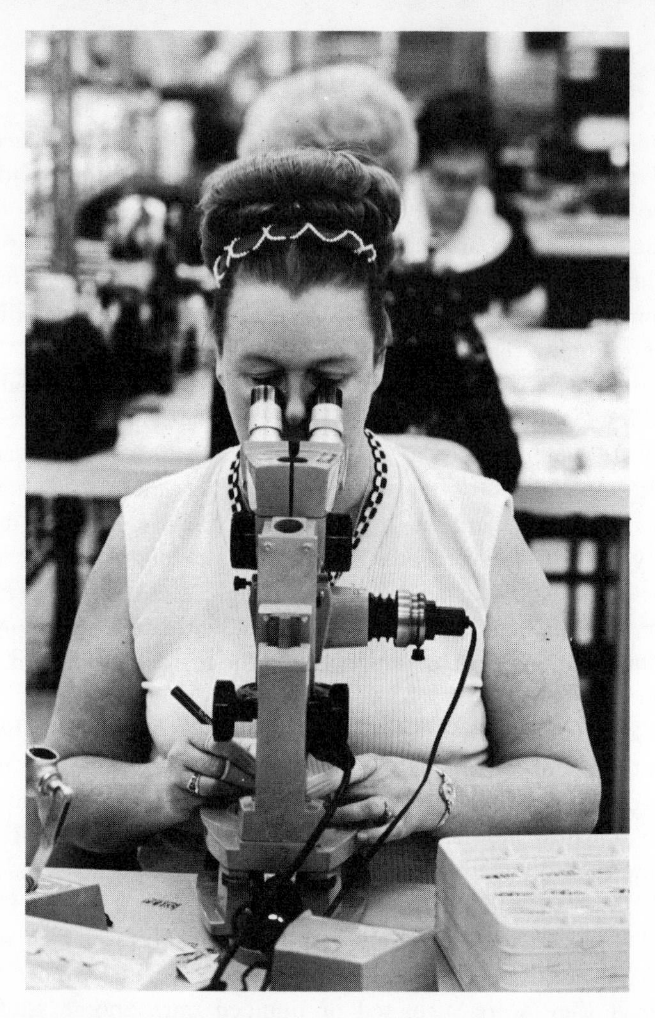

work wearing very nice clothes, clothes one doesn't usually associate with a factory. As Kay Cole said, "The girls who work in here don't wear cotton house dresses. They come to work nicely dressed." Another woman told me, "I wanted to find a job that wouldn't tire me out. I need some energy for my home and children." And Carmen Lingetti said, "I don't want a job that will make me get dirty. I don't want to do a man's job even if they make it available." Still, there are some women who wouldn't mind doing what other women consider physical jobs, and there are plenty of jobs that women might be anxious to do and that wouldn't require great amounts of physical work.

The company is pushing hard to get women to think about promotions. It has a plan of affirmative action to comply with the Equal Opportunity Employment Act, which requires certain numbers of women in certain places at certain times. The company is encouraging women in different ways. One of the most interesting that I heard of was a course given to twelve women to raise their work aspirations. Behind the program lies the belief that women have been conditioned to hold back and that with a little encouragement they would move ahead.

I asked several women about this. I got no clear picture, except that most didn't want to be the first in a job formerly considered to be all male; they preferred to wait and see what happened to the women who moved into those positions.

The company should make the necessary changes to promote real equal opportunity. However, it will make a mistake if it overlooks the potential problems, both at work and at home, that affirmative action may create. There is already a backlash developing on the shop floor because the company's openly stated policy of equal consideration is perceived as preferential treatment in promotion of women as well as minorities.

Many of the people who are most critical of the company's programs are men, whose potential for upward mobility within the company may be seriously reduced. Formerly, even when men entered jobs as 33 grade workers there was some feeling that with service and a good record there was a fairly good chance to move up in the company. Although there were large numbers of women, most were considered not interested in promotions, thus leaving a much smaller pool of people to compete against. So the best paying jobs in the shops and all levels of supervisory positions were, and still are, dominated by men. In the skilled crafts where there are 600 workers there are only two women, one of whom is a trainee. There are 378 section chiefs (first line supervisors) and only fifteen are women, and of the 137 department chiefs only one is a woman.

Women are now being encouraged to compete for these jobs. Competition for higher paying jobs has always been keen. Now many of the men realize that they will not only have to compete against one another, but that they will have to compete against women and minorities.

I found that, as a group, male layouts are particularly sensitive about the issue of female promotion. There are approximately three hundred layouts and three quarters of them are men. Many men view a layout's job as a steppingstone to higher employment in the shops and a possible move into supervisory positions. Now many of these men feel that things will be tougher, and that less qualified women will be moved up over them. They talk about this or that "dizzy broad" being promoted, or as John Gearen told me, "Hell, they'll keep promoting those women to look good to the government, and then we'll have to carry them."

Another thing that bothered many of the men was the fact that the company was dropping certain requirements from jobs so that women could enter what were previously considered all male jobs. Several tool and die makers said the company had seriously reduced the weight-lifting requirements for the jobs because women couldn't pass them. Karl Smith told me, "Not that I have anything against women. If a woman can qualify for the job the same way a man can, she should get it, but don't change the qualifications just to make it easier for the women."

There also may be problems because now that there are no jobs that are for men only, there are supposedly no jobs that are for women only. Men have begun coming in as 32 grade operators. Although men are currently given the more physical 32 grade tasks, as their numbers increase this will be increasingly difficult if not impossible to do. I don't know how numbers of men working on low grade benchwork will react to that work or how their presence on the bench will affect things.

Serious efforts to promote women may create problems at home for both men and women as well as at work. Many women are reluctant to make changes in work or move ahead if it will upset their home relationships. This is particularly true of the women who suggest that they have come to work to supplement a husband's income. Many of them acknowledge that their working bothers their husbands. Alice Smith told me, "I wanted to go to work, the kids were grown, and I didn't want to sit in the house. When I told my husband he said he didn't like the idea, that he made enough so that I didn't need to work. I told him, 'George, I want to.' He still said no. Then I told him we could get some of the things we always wanted if we both worked, and he finally agreed." Another woman told me, "My husband agreed to my working but when I said I was going to get a job as a waitress he nearly blew his stack. He said he didn't approve of his wife doing that kind of work. I couldn't argue with him so I came to work at Western, even though it meant making less money."

Women seemed very concerned about how their working made their husbands feel. "You know," Arlene said, "they talk about how

it's good that women are going to work, but they don't talk about what it does to a man. A woman working makes a man give up to a certain degree his sense of dominance, his feeling of 'I can provide.'"

Few women, I noticed, bragged about how much they made, and how much their husbands needed their contribution. Instead, most were quick to point out that they were supplementing their husband's income, even if they were making nearly as much, or even more.

What will happen when women begin to assume positions that will equal or surpass those of their husband's? I found quite a few women who were concerned about this. Maybe their husbands shouldn't be disturbed by the prospect, but many admitted that it would be threatening to their husbands and that they weren't interested in doing that.

I heard some women discussing a woman who had recently been promoted. "Did you hear," said Jane, "they moved Ann up to 211. It's pretty good, isn't it?" "Yeah," said Alice, "but I hear she and her husband are having a hard time." Later I asked Alice what she meant. She said she had heard that Ann's husband had a maintenance job with the town of Methuen. When Ann had come to work at Western she had started out on the benches. At that time she and her husband were supposedly getting along. However, since she had been moving up she's been hanging out with a different group of people, and now she makes more than her husband, and there's tension at home. Alice was one woman who strongly believed that women should support their husbands emotionally and was committed to very traditional ideas about the roles of men and women. It wasn't clear from her telling of the story whether economic conditions were causing tensions for Ann and her husband or whether that's how Alice needed to see the situation.

Affirmative action may have a major impact on many people who work at Western Electric. Jobs will undoubtedly open up for many women; whether or not large numbers of women will gravitate to the new openings is still a question. There certainly are some who need and want the chance to move up. However, whatever the new social values, most women I met at work, young and old, retained fairly traditional views of sex roles.

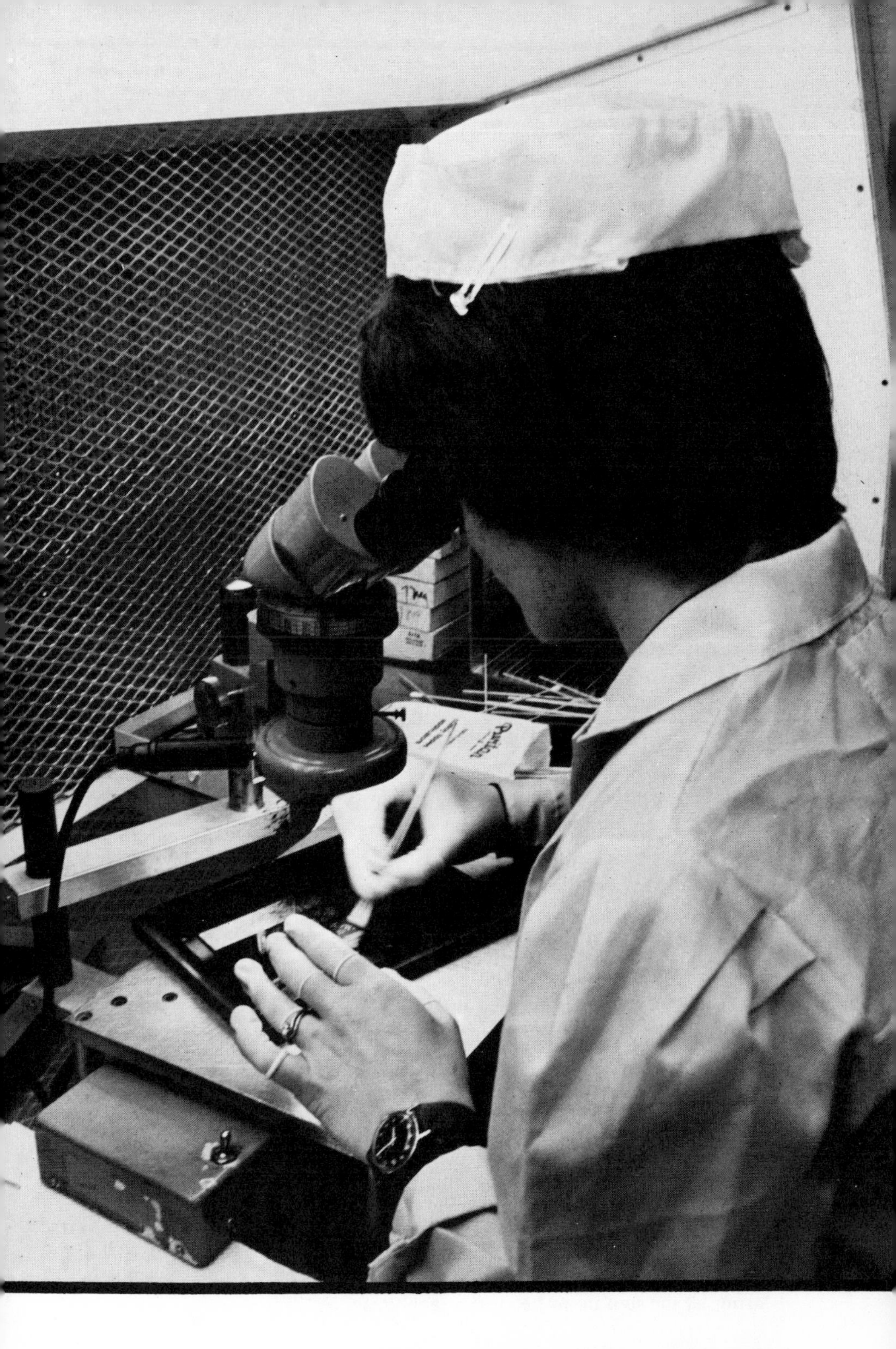

24. DAY AND NIGHT— NIGHT AND DAY

Rachael and Bob Couture—Rachael Couture often talked about the new house she was moving into. When I originally considered interviewing her, I planned to talk about the new house. It was only when I found out that her husband, Bob, also worked at Western, and on a different shift, that I changed focus and decided to talk to both of them.

"How was your day?"

"OK. How are the kids?"

"Fine. I started the wash. Will you finish it?"

"Sure. What's for supper?"

"Boiled dinner."

A typical conversation when a husband returns from work? Maybe, but these are the first few sentences of a ten-minute daily conversation between Bob and Rachael Couture as they sit in Western Electric's cafeteria. Bob, who has worked the first shift, is about to go home, and Rachael, who works the second shift, is about to start work. The ten minutes they share together between shifts is often the only time they spend with each other during the working week.

Bob, who is twenty-eight, and Rachael, who is twenty-seven, were high school sweethearts in Madawaska, Maine, where they grew up. Both are children of French-Canadian parents. Although they speak English with everyone else, they still speak French when they are alone.

"We met in high school," Rachael says. "We were pretty serious. We didn't talk about marriage until Bob went into the service after high school. I went to work at the Jade East factory ($1.40 an hour) and began saving. We got married after Bob got back."

"I was accepted at college," Bob says, "but my parents couldn't afford it. My father was a millwright at the paper mill. He had a second job and he was still bringing home maybe a hundred dollars a week, just enough to make ends meet with five kids. He couldn't swing putting me through college. I knew I had to get some schooling, so me and my buddy went downtown, saw an army recruiter, and the army let me sign up to become a technician.

"I signed up two weeks after I got out of school in June and I left in August. I spent most of my time over in Germany. I got to learn a trade, I matured a lot, and I saw most of Europe on government time.

"When I returned from the service I knew I'd have to leave town—there was no work. I moved down to my aunt's. I looked for a job in electronics but I couldn't find anything. I took a job at Western; it was the only place that had any openings at all. I took the first job I could get my hands on and then four months after I got in there I got a job as a technician, so I was all set.

"I was at Western for about two months when I went home to get married. I had to borrow money to even drive up and get married. At the time (1967) I was a 34 grade making $2.19 an hour.

"You know what our honeymoon was? We took four days coming home, and that was it. We started with nothing and now we're about to leave on a five-day vacation for Bermuda. Right now we're doing real well, we're making more than twenty-two thousand dollars a year between us. It makes a lot of things possible."

The extra money is to a large extent due to Rachael's returning to work. Rachael made the decision to start working a little more than two years ago.

"I wanted to get away from the everyday routine—regular housework—and I knew the extra income would help a lot. I enjoy getting out, being with different people. I went over to Western because Bob works there and it pays well. It's one of the best paying places around.

"I really like it. Sometimes I look forward to going into work. The people are very nice to work with, and the job isn't one where you get dirty. You can wear clean clothes to work and not get them all dirty."

"I didn't really want her to go back to work," Bob says. "We talked about it quite a bit. Rachael had worked before. When Andy was young she worked all over the place. She bounced round from one job to another, but after she got pregnant with the twins I kept her home for close to two years. It was tight. I had to start a second job part time again. I started my own little business nights, but that didn't pay anything. I didn't have the time to put into it. I thought about starting a regular second job again. I had worked for nearly three years on a second job doing regular factory work. I'd get out of Western, go home, have my supper, and then I'd go to work at Malden Mills at five and get home about midnight."

"I've got to give Bob a lot of credit," Rachael says. "I don't know if I could have done it [taken a job] without him because he really helps me with the kids and the housework. When I was going to start back to work we talked about what he'd do. He told me that he was going to help me. I didn't really think he would do that much. I figured he would help out for a week, do the things I'd do, and then he would say forget it. But no, he really has kept it up. He's a big help.

"He gives the kids their baths every night, which he didn't do before. He's real good as far as keeping up the house. If I haven't done something, I don't even have to tell him, he'll just go ahead and do it. He washes the dishes at night. He even helps me with the laundry and vacuums the rug if I haven't had the time to do it during the day. I get the supper ready before I leave. He feeds the kids. He's good with them. You have to take into consideration getting stuck with three kids is no easy thing. He's got more patience than I do with them, really. I guess what it is is that he understands that I'm helping him, so he's got to help me too."

"I do a good portion of the housework," Bob says. "Like I helped her with the wash tonight. I'll vacuum and clean up the house for her. I told Rachael from the beginning that I'd help her out. I don't know, I don't think she really thought I would. She had been doing most of the housework, but with her working I figured I should pitch in. It really wasn't that hard for me, being the oldest in a family of five, I was used to it. Plus being in the service I learned to wash my own clothes, do my own ironing, clean my own room. I don't really mind. I know I have to do it because if I left it all on her it would be ridiculous. She'd go nuts. I just have to help her out, that's the way I feel. If I didn't help her out I wouldn't be holding up my end of the bargain. I think most men whose wives work feel that way."

"Working is something I like to do," says Rachael. "I do wish Bob and I could work the same shift, but with small children it isn't right. I took a job on the second shift because I wanted to be home with the children. When they're so young they need their parents. Also if I had to have a baby-sitter for three kids it probably wouldn't be worth my while at all. We've worked out a pretty good system. I'm with the kids during the day and Bob takes care of them at night.

"I take the kids to the day care center on my way to work. Bob picks them up on the way home, so they're only there for an hour a day. That school is good for them, too, because they get to play with other kids. They teach them the alphabet. The twins, they're not even three years old and they come home with their papers and they really like it.

"I do have to make some adjustments at home, so does Bob. He gets up at a quarter of six and goes to work. The kids and I get up around seven-thirty. Since I don't usually go to sleep until twelve-thirty or one it doesn't give me that much time. By the time I give the kids their breakfast and get them dressed and do my daily cleaning it's time to make dinner for the night and feed them lunch. A couple of hours later I've got to get ready to take them to the day care center and get ready for work.

"I've also had to change some of my housework routine around. Whenever possible I try to do the laundry and the ironing on Sunday nights now. I have to do the grocery shopping for the week on Saturday when I used to do it on Thursday or Friday. The only other thing is bigger jobs, like doing the floors; I have to do them on the weekends.

"The real disadvantage of it is that it cuts down on the time I get to spend with Bob. I see him at work for about ten or fifteen minutes and we have coffee together before he leaves. He calls me up three times a day and at night I call him up. Really, we have weekends—that's about all.

"The only other time I get to see him during the week is late at night. I usually get home about a quarter to twelve. Sometimes Bob will have dozed off and when I get home he wakes up and we chat for a while."

"I don't wait up for her every night," Bob says, "but quite a few nights I do. Sometimes I'm not that talkative. When she comes in, she's still wound up from the night, and she expects me to be wide awake and sit there and talk to her, and I can't do that, and she doesn't understand that. I try to talk but my eyes start closing. She's getting used to it now, it used to bug her. She'd be talking to me and I'd be falling asleep."

"I'd say," Rachael says, "with all the disadvantages, the advantages still outweigh the disadvantages. It's made it much better in terms of money. We could have gotten by if I didn't get a job, but we wouldn't have anything extra. This way now we're really living. Like we're going to Bermuda on vacation. We're going for five days during vacation. This woman who works with Bob is going to come over and baby-sit for the kids. Never mind that it's my first trip to Bermuda, it's my first flight ever. I've never been on a plane before. I know darn well if I hadn't been working we couldn't have a lot of extras. We just put up a fence around the backyard. It cost us two thousand dollars. We got two brand new cars. We are living very well."

Bob says, "If there was just my pay we wouldn't be able to go out weekends. We wouldn't be able to buy lunches. I'd probably have to get rid of one car. Little things like that. We wouldn't be able to go away as often as we do. Like now, once a month we go down to Connecticut to see my parents. We'd have to cut that out with the price of gas. We wouldn't be able to go up to the beach whenever we wanted to. We wouldn't be able to go clubbing. We would just be basically home folks–stay in the house all the time. We wouldn't be able to do anything.

"You know, though, if I could make an extra five thousand dollars on a part-time second job, I'd still like Rachael to be at home, even though we wouldn't have so many extras. I say this, but one of the problems is I can't keep her home. She's not a homebody. When we came down from Maine we were only down two days when she got a job. I come home from work and she told me she got a job. You just can't keep her tied down. She's worked hard all her life so she'll always want to go to work, but I still haven't changed my mind. I still hold the same ideas about women working. She really should stay home with the kids, but the way it costs to live today we couldn't enjoy a decent life style if she didn't work.

"My salary has gone up from when I started with Western at $2.19 an hour to $6.17, but as much as we keep making, everything else has gone up. This last raise, three years ago, we got about seventy-one cents an hour. That seventy-one cents got eaten up [he snaps his fingers] like that.

"Everything just keeps getting more expensive, taxes on the house, everything is skyrocketing. I can't believe it, our grocery bill has almost doubled in the past two years. Of course the twins are growing up, but that's no reason for them to double. Rachael uses her paycheck, which is about fifty-nine dollars a week after savings and taxes for groceries and gas for her car, and she hasn't got any money left.

"I'm not complaining, we do save. Rachael puts thirty dollars away a week and I put away seventy-five dollars and we live on the rest. I like it being taken out of my pay. I'm the type of person when I have the money, and there's something I want, I'll go out and buy it. Whereas, if it's taken out of my pay each week, I won't notice it. Then when we want to do something it's there. We used that money in the credit union to go to Bermuda, we used it to buy the second car, we used it to buy the fence, plus we used the credit union to finance the other car.

"Economically, we are doing real well. However, I'm not saying there aren't any problems to us working two shifts. It's hard on both Rachael and me, and it's hard on the kids. They miss having her home. For example, quite a few times they've asked me when's Mommy going to come home. They want to stay up and see her. Now they're starting to understand that she's working nights and they can only see us together on weekends when both us aren't working. A few times it's been especially hard, but we've worked it out. My son especially missed her when she started working because he was older and the twins were young enough so it really didn't make a difference. My son, he's attached to both of us, and he was spoiled rotten, and he missed Rachael quite a bit when she started working. Now, maybe once every two weeks, he'll ask me when she's coming home.

"On the other hand, since Rachael's begun working he sees a heck of a lot more of me. Andy enjoys that because I take him to softball games. I do more things with him than before. There isn't a doubt that Rachael's working has made life easier for me. I'm more relaxed now that I don't have to work two shifts. I have a chance to play softball. I have a chance to go to school in the fall.

"I don't think it's hurt our relationship. Actually I think it's helped it. When I was the only one working Rachael was cooped up in the house all day with the kids and the housework, and sometimes she'd get irritable. We argued much more then. Rachael's not the kind of person you can coop up in a house.

"Since she started working we argue less, it's 500 per cent better. As a married couple you see each other every day, and tend to argue. We hardly see each other and we rarely argue now. We might once in a while, but not often. We really don't see each other until the weekends and we have a wonderful time together.

"I'd like Rachael to go back on days eventually. Not right now because the kids are still too young. She has to get them ready for school in the morning. When they're about ten, then they're old enough to get dressed by themselves and go to school. When you start work at six-thirty it means you'd be out of the house too early and they'd be alone for too long a period."

"I'd like to keep working," Rachael says. "I almost quit last November. The kids needed me at home, but then when I started thinking about the good job I have and all the money that's coming in the house, I decided to stay on. Now that I'm used to working I think I'd find the evenings boring. Not that I find my husband boring, but I can't get interested in TV anymore because I'm so used to not watching it. Even on the weekends, Sunday nights, I can't watch it. I used to watch a lot of TV. I'd do my housework during the day, but at night I'd almost automatically turn the TV on.

"This way Bob can go to school too. He goes to school three nights a week. He's studying for his B.A. He's only gone two years. He's got to go about, I don't know, five or six years before he'll get his degree."

"I've decided to try to get a B.A.," Bob says. "If I were to get an associate's degree in electronics, I wouldn't make any more money than I'm making now. I know two fellows, one of them left Western six years ago, he went into engineering and just this year he got a raise and he's making the same money I am. All of these years, he's been making less. When we get our raise at the end of this month, he'll be making less money again. He's got more responsibility than we do; he's answerable to more people and he's making less money. So there's no incentive there. Right now I'm taking data processing and business administration. If nothing pans out at Western after I have my B.A., I'm just going to go looking somewhere else.

"I'll probably find something somewhere else faster than I'll move up at Western because of the way they operate. That's the bad part about it. If you don't have a godfather it's really tough to go from the shop to supervision, especially now when they're giving preference to women."

Standing on the Couture's ranch-house porch, Rachael watches the children playing in the large back yard below, secure that they won't get lost because the new wood fence is up. "We really like it here," she says. "We like this house—it reminds us of our home town."

25. TWO SHIFTS

The majority of people in the shops work the first shift. However, there are second and third shifts on which several thousand people work. After working on the day shift for four months I got a chance to work on the second shift.

Ed Patrick,[1] who had correctly predicted I'd be sent to work in one of the thin film rooms, said, "See, what did I tell you? After you've sweated down here they want to put you on the cozy job before you leave. [We had just been through a very tough summer, the last one in the shops without air conditioning. The temperature occasionally rose into the high nineties, making work very uncomfortable.] You won't have to worry about the heat anymore, you'll be in a climate-controlled environment."

When other people learned of my new assignment, they wanted to know who my supervisor was going to be. I didn't know. People told me to find out, because who your supervisor is is as important as what your job is going to be. Ann Bourque told me, "There are some supervisors in here I wouldn't work for no matter what they offered me. You better hope you have someone as good as Jack."

I took a lot of kidding about the new job before I left the first shift. "It'll be like a country club upstairs," Debbie Harris said. "Hell," Larry Carney said, "you'll think you're on vacation up there."

"You'll see," Andrea Hoeffler told me, "they work differently on the second and third shifts when the big shots aren't around. It's much more relaxed. I think you'll like it."

Although I only worked upstairs on the second shift for a month, I discovered that there were many differences between the shifts, the tasks, the groups, and the environment in which people worked.

Unlike the equipment shops, which are open and in which workers are required to wear no other special clothing than safety glasses, the thin film rooms are enclosed and climatically controlled spaces. White smocks and white caps are mandatory wearing apparel. Where I worked most people were also required to wear "finger rubbers" to protect the work from grease.

[1] The names in this chapter have been changed.

One woman told me that working in the climatically controlled thin film rooms and wearing a white smock made her feel as though she was working in some fancy place, such as a hospital. She said, "I know I work in a factory, but when I talk about a factory I'm really talking about the shops, and not the thin film rooms."

There were some immediately noticeable differences between the tasks performed in thin film and in the equipment shops. Most of the work in the equipment shops is done by hand—hand inserting parts, clipping leads, and soldering those leads. The work done in thin film rooms is primarily done by machines. Physical labor is minimal. Most of the work is tending machines, keeping thin film circuits moving through a variety of processes.

Most people I met at night seemed to like the predominantly nonphysical work we did. Many who had worked in coil winding or at hand inserting were glad they had been surplused or transferred into thin film, but I didn't particularly like the work. I discovered I was much more bored at night than in the equipment shops. The work was easier, as people had said it would be. However, I found that my job at night was primarily as an attendant to a machine. I had a lot of free time on my hands. There were often long periods during which I was waiting for a machine to finish a process. I could, and did, move around, talk, and read while waiting for the machine. All the nonbusy time made the evening go very slowly. I realized I preferred being busy as I had been most of the time downstairs.

Although I didn't particularly like the work we did, I found that I much preferred working the second shift hours to those of the first shift. The second shift is not supposed to be the attractive shift. In fact, the company pays a 10 per cent bonus to those people working on it. New workers are usually assigned to the second or third shift. There are some people who work the shift by choice, either because of child care needs or time preference.

Several people told me they liked working the second shift more especially in the winter because it meant they didn't have to get up at four-thirty to shovel their driveways and to come to work in pitch dark. As mentioned, I also preferred the second shift's hours. I never got used to getting up at 5:15 in the morning. Even though the second shift didn't get out until 11:30 P.M. and eliminated most social life during the week, it saved me from getting up very early, which is something I deeply appreciated.

I soon discovered that the major difference between the shifts was much more of a feeling of freedom on the second shift. For a while I thought this feeling was the result of different supervisory styles. The two men I worked for were quite different.

Harry Dresden, my second shift supervisor, is a college graduate holding an engineering degree. He is what many people at Western consider the new breed of first line supervisor. He didn't work his way up in the shops into a supervisory position. Instead, he came over to the shops from the offices. He is a warm, open person who is not reluctant to talk about himself, his family or life. Harry has a good reputation with the people who work for him. As one worker told me, "Harry doesn't care so much about those diddly-shit rules: be here, do this, do that—as long as you do your work, and don't take advantage of him."

Jack Colpepper, my supervisor on days, has worked for the company for twenty-three years. He received no college education and worked his way up through the shops. He is part of the old guard. Jack, unlike Harry, is very closemouthed about his personal life. Jack has a reputation as a good supervisor who is particularly effective at starting up new jobs. He is known as a company man, a person who lives and works by the book. He is strict but fair. Without exception I was told by people who had worked in other shops that Jack was one of the best supervisors to work for.

These two men brought very different approaches to their jobs, and I'm sure that if they were working right next to each other they would treat their groups differently. Nevertheless, both were rated by people as good supervisors.

I didn't spend time working for a bad supervisor, but I heard a good deal about them. A bad supervisor keeps his people under constant pressure. I heard about supervisors who wouldn't let people out of their seats. A couple of women told me they had worked for supervisors who made them so nervous that they developed ulcers. I also learned that a bad supervisor was one who didn't care about his people. No one defined what that meant. Good supervisors respect their employees, although they may bring very different styles of supervision to their groups. Bad supervisors don't.

By the time I stopped working on the second shift, I realized that the greater freedom of the second shift was due less to the differences between individual supervisors than to presence or lack of presence of higher-ups in the shops.

Before 5:00 P.M. the second shift and the first shift seemed very much alike. If you wanted to take a break, if you had nothing to do, you left the area and went to the bathroom to smoke, talk, or read.

After five o'clock, things changed dramatically. People actually read near their work. They walked around the area and spent more time talking with one another. They changed the way they operated the machines. This transformation was primarily due, I learned, to the fact that most of the managers, engineers, and "big cheeses" from Mahogany Row (the top floor management) had gone home at 4:45.

On the very first day I began working upstairs I saw someone, Donald Jones, reading a paper near his work. I was flabbergasted. I couldn't imagine anyone sitting and reading by his work downstairs during the day. If you wanted to read you left the area. I asked Donald about it and he said, "As long as you're discreet about it, and do it after the bosses leave, Harry won't hassle you. Just do your work, and you can read when there's nothing to do."

Many other things changed after five o'clock. We didn't run the machines at the same rate in the first part of the evening as later on. One day Jerry Kelly told me, "After quarter to five when the engineers go home, we do things up here the way we want."

That was exactly what happened. Before five the developing and laminating machine that I helped two people operate was run exactly the way the engineers wanted it to be run. The machine required a feeder, who fed the circuits into a conveyor belt. The circuits then went through a series of sprays and were removed from the belt by a catcher at the other end. The engineers, who were frequently walking around before five, wanted the machine operated at a belt speed of between two and three. That was how it was run until they left. Then the machine was often sped up. Although the speed was left at the lower numbers during the first few developing stages, when it came to the last spraying we often sped the machine to six or seven, for no other reason than it was more challenging to try to catch the boards without them backing up at the higher speeds. Occasionally when things were going slowly and we were bored, we had a contest to see just how fast we could run the machine without circuits backing up.

Although on its surface this might sound reckless, the fellows working the machines seemed to know just how fast the circuits could be run without their being spoiled for inspection. They also had their own ways of mixing chemicals and measuring things. I was amazed at the little tricks and the ways tasks could successfully be sped up when people wanted to work faster.

The more relaxed environment on the second shift carried over from the work into the way breaks, meals, and leaving time were handled. On the second shift we were never pressed to get back from a break exactly on time. Many people, in fact, often left just before the break was to begin and returned a minute or two after it had finished. As for dinner, most people clocked back in before the half hour was up, and sat outside for a last smoke or some conversation, often not going back into the thin film room until a few minutes after the bell had rung.

People generally quit much closer to quitting time on days than nights. I found that, although people might slow down after 2:30 on days, they rarely left their seats to get ready to leave until five minutes before the end of the day. People didn't seem to feel that compulsion upstairs. One couldn't ignore the time clock, but it didn't seem as controlling on nights as days.

I told a friend of mine who worked on days how much freer things seemed upstairs at night. Ed Patrick wasn't surprised, nor were the other people I talked to on days. What they said verified my own feelings. Supervisors on days get a lot more pressure from higher-ups than supervisors on nights. As Ed Patrick said, "On the first shift

supervisors can't afford to have someone standing around doing nothing, because they never know who may be walking around watching."

First shift supervisors, knowing that higher-ups may be watching, are very concerned about how their area and people look. They tend to pass the unstated pressure they feel onto the people working for them. When the higher-ups leave, second and third shift supervisors receive less of this unstated pressure and therefore are more relaxed with their people.

Many of the people I met on the second shift had worked at some time on the first shift. To a person they preferred the less pressured environment of the second shift. Bill Grego told me, "Sure I like it better. There's nobody snooping around watching you all the time. After five we don't have to worry about the bosses walking in and out. There's just not all the pressure and hassle that you've got on the first shift."

CAUTION

I wondered if the greater degree of informality on the second shift meant that less work was accomplished. The people on the second shift didn't think they worked any less hard than people on days. In fact, several claimed that even with—or because of—the looser atmosphere they produced more than people on days. The fellows who ran the machine I worked on were always complaining about how much more they got done in an evening than the people working on days.

I asked them, and others, why they thought they got more done. People said they liked to know in advance how much work had to be done and then be given the freedom to choose how to do it. "Look," Irene Landers told me, "you pace yourself whether they're watching you or not. This way up here we know what we have to do, and we do it. That's how work should be."

The more I thought about the way the work was done, the more I realized that people on the second shift did seem to have a greater amount of freedom in deciding how to schedule their work. While working on the first shift I felt that no matter how much I did there would always be more work around. In fact, if there wasn't work there would be "make work," so I kept myself looking busy. Aware that as long as they remained in their work area they were expected to work, most people had internalized their own pacing system. This constraint was not nearly as strongly felt at night.

I was glad that I had an opportunity to work on the second shift for a very brief time. It gave me a chance to see a different group, work on different types of jobs, and meet new people.

Unless people at night produce less, and they don't appear to from what I could learn from the management, there seems to be little reason why the looser, less pressured, more open atmosphere of the second shift should not be exported to the first shift. Control over output, which is the worker's covert prerogative even when it is not his overt privilege, is an important element of job satisfaction. By allowing the second shift worker to organize his own workday rhythm even unintentionally Western Electric allows him a valuable sense of daily accomplishment which is being denied to the majority of first shift workers.

CONCLUSION

In spite of the huge economic contribution made by the twenty-eight million men and women who work in its factories, American society tends to ignore or to denigrate factory work and blue collar workers. The view that people who work in factories are somehow failures, entitled to relatively low social status, is pervasive. For example, a series of public service TV commercials, aimed at trying to convince drop outs to finish school, made the point that unless one finished school one would only end up punching a time clock in a factory in a dead end job. These nationally broadcast messages certainly conveyed the most negative view of factory work.

This view is buttressed by the organization of plants. Through such organizations, no matter what a company intends, workers are made to feel unimportant. Company bulletins or national commission pronouncements on the importance of what workers do cannot alter the basic reality that daily convinces them of the contrary. The message most effectively conveyed is that workers are exchangeable, replaceable components.

There is much that can be done to alter the existing conditions of the shop floor, to change the way jobs are done and the atmosphere in which work is accomplished. I promised to share what I learned at Western Electric with the local management in the hopes that such discussion would lead to changes within the plant. I made numerous suggestions, including changes in the orientation and training program, the removal of quality control charts and time clocks, permission to smoke wherever possible, a better explanation of the bonus system, less authoritarian roles for supervisors, more freedom for shop personnel to organize their work, and a much greater role in decision-making by shop employees. So far the only specific change resulting from my work has been in the area of quality control. The charts have been experimentally removed from stands in several departments including the one in which I worked. Mistakes are still charted, but they are no longer publicly displayed. Instead, the supervisor keeps them in his drawer. Still, Dave Hilder, general manager

of the Works, has stated that my observations have helped "make a good operation better," and have made him and those around him more receptive to suggestions for change on the shop floor.

An experiment, a motivational enrichment trial (MET), was begun in January of 1974 in one of the shops after I stopped working. The experimental unit is commonly referred to throughout the plant as the "fishtank" department. It got this nickname because the workers were allowed to bring a fish tank into the department. Most workers I talked to around the plant knew little about the MET program aside from the fact that radios, plants, and fish tanks, prohibited elsewhere, were permitted in the department.

These seemingly small changes proved to be very important. They not only have made the work environment more pleasant for the people in the MET department, but they are also a tangible sign of privilege, of being allowed to do things that other shop employees cannot do. The notion of one group, namely the office workers, having privileges denied others is one of the biggest irritants I found on the shop floor. Punching a time clock, never enjoyable, is more oppressive when the office help doesn't have to do it. Not smoking is a hardship for a lot of people. Not being allowed to smoke in the shops and thus having to sneak into the john, while someone else in an office smokes openly, is infuriating. It is not surprising, therefore, that there was almost an immediate demand in other parts of the plant for permission to bring radios and plants into the shops.

Beyond such superficial signs of change, those running the MET trial think there is a better, more human way to run the shops, which are currently organized like classrooms. This organization reflects a presumption that workers are childlike—irresponsible, incompetent, and unable to make decisions. The teacher has been replaced by the supervisor. Authority flows from the top down. Workers are all too frequently made to feel that they are constantly being watched. I was conscious of these feelings both from talking with workers and from working on two shifts. I learned from working on the second shift, where there was more freedom and a more relaxed atmosphere, that work can be arranged in a less authoritarian manner.

The organizers of the MET program are moving in that direction, and hope through a variety of techniques to move supervisors out of what they aptly describe as the "police punishment style of leadership" into a more nurturing role. What's more, they want to change jobs and involve workers in decisions about their work. There are some worker advisory committees within the department. Some efforts have been made in the area of job restructuring including job enrichment, job enlargement, job rotation, and teamwork.

I talked to some people working in the experimental department. I thought that they would talk about changes in their tasks or about the work committees, but what dominated our conversations were remarks about the environment and atmosphere in the department.

One woman I did not know said, with a note of embarrassment in her voice, "This might seem crazy to you, but I look forward to coming to work now. Really. Last week my husband said I hadn't taken a sick day in a long time, asked me why didn't I take one. But I told him I didn't want to. I like work. I don't feel any *pressure* anymore. I enjoy going to work, and being with people, it's just much nicer."

Another woman said, "The only thing I worry about is if I get transferred to another department where it was like it used to be, I don't think I could stand it." None of them talked much about the actual work even when pressed to do so.

These comments have made me reevaluate many things I've heard at work. For a long time I interpreted negative, work-related comments to indicate how alienating the jobs themselves were. Now I understand that when people appear to be talking about their jobs, they are frequently talking about the environment in which they work, and the way they are made to feel about the work they do.

Before I went to work on the second shift, many workers told me how much I would like the job. No one ever mentioned that the work would be interesting. Instead I was told I'd enjoy it more because there was more freedom on the second shift, the room was air-conditioned, and the environment cleaner. As it turned out I didn't like the work on the second shift. Everything I had been told was true, but the work was more repetitive and routine. Several people on the day shift thought I was crazy and didn't know a good job when I had one.

When people talked about more attractive jobs they were focusing on a whole set of freedoms, amenities, and privileges enjoyed by other sorts of workers, not more satisfying tasks. How much more interesting *are* the tasks performed by most non-blue collar workers? It is the atmosphere that surrounds them, and the control that people have of their work, that makes many other jobs more attractive.

Those running the MET trial believe they have to do more than change the environment to make work more enjoyable and rewarding. They hope to restructure jobs by making them less fragmented and repetitive in an effort to make work more satisfying and less alienating. They'd like to enlarge jobs, to have people work on more of the product. When the fragmentation is an economic necessity, they hope to rotate jobs, so a person isn't stuck doing one job. They'd also like to promote teamwork, allowing groups of workers to decide how they want to break a job down.

Not everyone, however, wants a challenging, enlarged, or enriched job. Many people don't come to work in search of job satisfaction but because of the money. They want to invest only a limited amount of themselves in work. Some of these workers prefer, or at least don't mind, rather routine tasks that allow them the freedom to think about other things or to be more "sociable."

There are other workers who derive a sense of security from knowing the boundaries of their responsibilities, and who are not anxious to give up their routines for unknown enlarged jobs. Others find satisfaction in knowing how to do their jobs well. Although many of the tasks on the shop floor are rudimentary and look easy to learn, I found that many of them take a long time to do well. It may not take long to learn the procedure, but to be able to do it efficiently takes some time and skill. There are subtle differences between knowing a job and mastering it. Many times I'd do something as "required" only to be told by a more experienced worker that it was sloppy, and it wasn't done the right way—it wasn't done just so.

Still, there are plenty of workers who don't feel this way, who feel trapped in jobs that they consider too small or too limited. Many would be anxious to participate in and benefit from the type of restructuring that the people running the MET trial are attempting, but are skeptical about how much can actually be restructured. I can remember sharing a paper I had read about job rotation. After reading it, one remarked ruefully, "Terrific, they're going to give me ten boring jobs to do instead of one."

Among workers the greatest resistance to changing jobs is founded on a basic distrust of management efforts. Many workers believe that

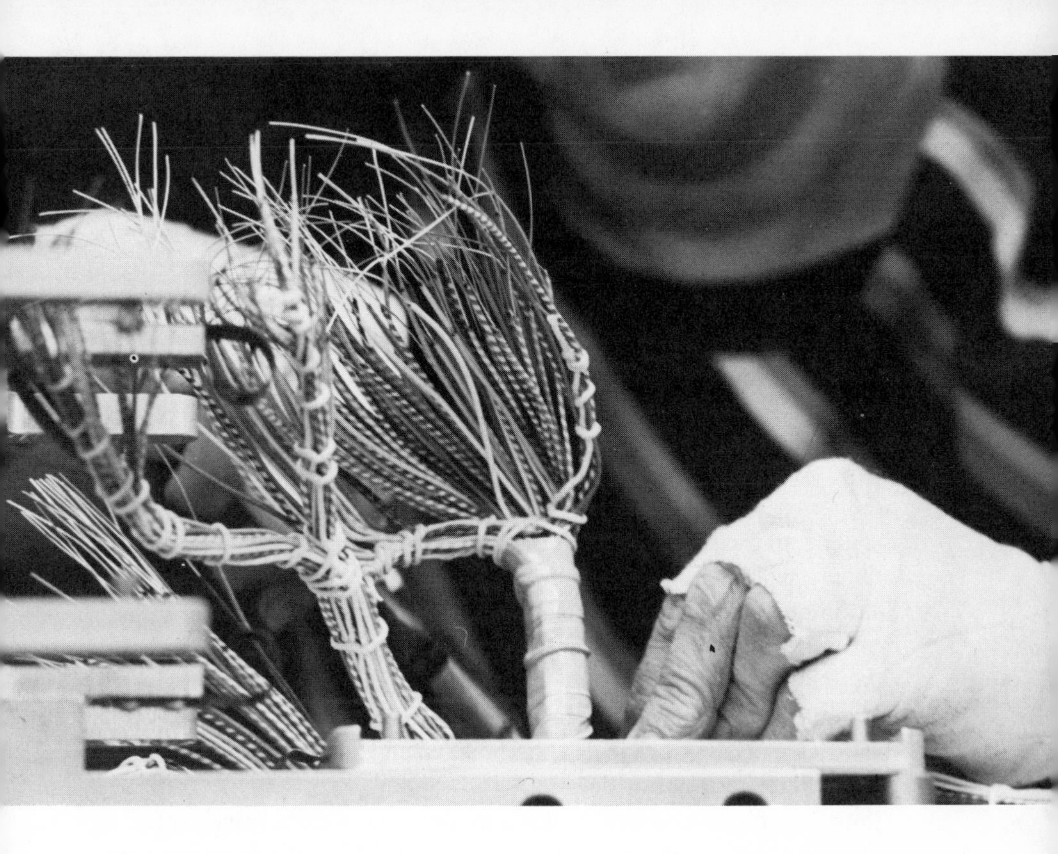

the company's interest in reshaping their jobs has less to do with their job satisfaction level than with increased productivity.

One day I was talking about the MET program and a worker said, "This company is not in business to make people happy but to make money. If they make money and people are happy so be it." Although the people running the MET program believe they have the commitment from top management to "humanize" the work environment, and restructure jobs because it's *right*, they do justify what they're doing in terms of increasing productivity. An MET memorandum states, "It was agreed that the attainment of effective production levels should be the main objective."

Joe Miett, department chief of the D2 D3 Channel Unit Shop, defends this position as necessary to sell the program. Sell the program to whom? Certainly not to workers, so it must be some segment of management. Underlying such thinking must be the assumption that factory workers are paid to produce, not to enjoy or be satisfied by work, and that they must *earn* a more enjoyable environment and work organization through increased productivity.

I am certain that workers could increase production if they wanted to. Workers are ingenious at finding short cuts to beat rates set by production engineers. Factory workers, not surprisingly, know a great deal about their own jobs. They have a reservoir of knowledge that is underutilized, since little in the current work structure encourages workers to share their knowledge. There is some sharing among workers but the knowledge is usually withheld from management. Management is aware of this and hopes that instituting changes in the environment and jobs will make workers more receptive to sharing what they know. My guess is that there will be some immediate increases in productivity because of MET, but over a long period, if management is calling all the shots, the increases will not continue.

There will be a continuing demand among workers for greater freedom in the organization of work, and for work restructuring. These demands will join but not replace more traditional demands for security, fringe benefits, and pay increases, which have long dominated the concerns of workers. The question that remains is not whether these demands will be made, but how changes in the work environment will be accomplished.

Small cosmetic changes—radios and fish tanks—are relatively easy to make, and can be instituted by management fiat. What about more basic changes? Many workers don't see the possibility of beneficial change occurring outside the fundamentally adversary relationship between management and labor. Is it possible for the prevalent *us* and *them* atmosphere at work to be replaced by a feeling of *we?* Is conflict between management and labor an inherent element of capitalism? Is there a fundamental conflict between the human needs of workers and the profit needs of business?

There are no easy answers to such questions, but if there is any possibility of genuine co-operation it will not come out of a system initiated, planned, and directed by management alone, because management and workers often have different and conflicting interests. Even if both sides were truly represented, it isn't clear that the problems could be worked out. Without real representation of both sides, nothing more than inadequate, piecemeal improvement can occur.

If one looks at what is being tried at the Merrimack Valley Works in the time perspective of a year or two, the changes envisioned seem substantial. However, if one looks at the changes being discussed at Western Electric and elsewhere and realizes that these changes have

SAFETY
STORE

their antecedents in experiments in the 1920s at Western Electric's Hawthorne plant, one must be skeptical about the pace of change. Those experiments revealed that, almost regardless of how the environment was tampered with, workers produced more when they were made to feel important. Much of what was learned at Hawthorne a half of a century ago has been ignored and forgotten, although it periodically re-emerges, repackaged and relabeled, as a new way to make workers feel better about what they do.

For permanent changes to occur on the shop floor it is necessary for workers to participate in making the decisions that affect them. What is needed to reshape the workaday world is a form of work democracy in which workers and management jointly plan goals and operating procedures. This probably will not happen until and unless attitudes about factory work and the people who do it change.

Going to work at Western Electric meant getting to know people. Many of those peoples' lives have changed since I left, as has my own. Irene Lambert has moved her family to a new house; so has Carol Leavitt. Vicky and Harry Wrigley were robbed. Many of their belongings and some furniture were stolen, including some irreplaceable things that had been with them since their marriage. Vicky feels much more vulnerable because of the robbery. Buddy Barnes spent a month out of work recovering from an industrial accident where he nearly lost a finger. Mike Beal almost poked out his eye with a pencil in a freak accident at work. His vision has been permanently impaired, though only slightly, and he is now recovering from the injury. Bill Beal and his wife, Carmie, spent their first New Year's Eve in eighteen years away from their house, across the street at a neighbor's. Rachael Couture has had stomach problems, and her doctor has advised her to relax more before she develops an ulcer. Bucky and Alba Bocuzzo took a car and small truck and have moved themselves and their belongings to Florida. Arthur Barnes is back at school taking courses, and Wayne Barnes has taken a job in a hospital in another part of the state. She comes home nearly every weekend.

The current recession has affected a number of people. Harry Wrigley, who has sold cars for many years, is out of work. Pat Collins, who left the security of eighteen years at Western to sell used cars, was happier but couldn't make any money. He left that job and now is trying to sell other things. Nathan Lambert is helping his mother with the bills through his job as a drummer. The recession has taken a heavy toll in the plant. After nearly sixteen years as a layout, John Gearen has been "bumped" from his 36 grade layout job to a 34 grade job. Greta Barnes has kept her grade but has been "surplused" into another department. Layoffs began in the last month of 1974. By the end of May 1975 more than 1500 hourly employees had been laid off. Nearly everyone with less than five years of service was laid off. One month after Gandolfo Cascio received an award for twenty-one months of perfect attendance, he was laid off. It wasn't long before his brothers Matteo and Joe joined him among the ranks of the unemployed. Junior Valentin, Rachael Couture, Irene Collins and her daughter Joyce have also been laid off. Most of these people are waiting and hoping to be called back.

I learned much more from working at Western Electric than I had anticipated. Several friends asked how I could tolerate what they considered the boring, repetitive nature of factory work. I felt somewhat uncomfortable telling them that although I probably would have felt differently if faced with ten years of benchwork, I didn't mind it for five months. It gave a solid routine to my life and that felt comfortable. I looked forward to going to work most mornings, not because of the work, but because I had friends at work. Working at the plant reinforced my realization of how lonely writing is. The typewriter is not very good company, nor are the walls I often look at. I enjoyed the social community of which I was a part.

Initially I entered this community as an outsider. Over time people accepted me and shared their lives with me. It was not a one-way relationship. I came to share much of my life with them as well, and learned important things about myself and my family from the experience. Talking with Vicky Wrigley about her difficulties adjusting to new social values and her fears for the security of her children, I found myself telling her about my own mother's fears. Bill Beal and I discussed his problems with his son. I ended up sharing with him my not too dissimilar stories about problems I had had with my father. Discussing with Mike Beal his need for independence meant telling him of my own needs for independence and how I had tried to handle them.

Those conversations have pushed me out of my comfortable, self-satisfied sense of things—of knowing my parents. For example, during much of my adult life I have been in painful conflict with them.

We have disagreed, often bitterly, about many decisions concerning my life. I didn't really understand my parents. Greta Barnes brought that realization home to me one day after a long family problem discussion when I both wanted and expected her sympathy. She said, "You were really spoiled. I can see how upset you made your parents."

I'm sure she was right. I didn't understand what was important to my parents. How could I when I knew almost nothing about my parents' backgrounds, about our family's history? My parents rarely talked about their childhoods. When they did it was usually to remind us how far they had to walk to school or what they had to do without. I never visited their old neighborhoods, though both of them grew up in different parts of New York less than thirty miles from where we lived.

My parents gave me love and a relatively happy childhood, but never gave me a sense of where they came from, a sense of roots, a sense of who they used to be. I used to think this was just a quirk on their part. Now I think it is a much more general problem, an unattractive side of the American dream. For many people like my parents who lived through the Depression, the American dream held out a way to improve on modest beginnings through education, hard work, and monetary success. My parents took part in a post-World War II prosperity in which large groups of people moved from old city neighborhoods to the new suburbs. Unfortunately, getting ahead for many people meant leaving a lot behind. My parents wanted for me and my brother and sister all the advantages they didn't have. In trying to give us those material advantages they sacrificed giving us a knowledge of the past.

I found some of that past in the lives of people at work. My parents' backgrounds are rooted in economic circumstances not dissimilar from those of the people at Western Electric. A couple of women I met were, under a superficial layer of differences, very much like my mother. I found that as I got to know them better I was getting to know my mother better. When people talked about their thwarted dreams and ambitions, about their hopes for their children, I could hear my parents talking. Learning about my family and myself has not been painless, but it has been inestimably important. It has given me a better sense of who I am.

There were other things to learn from the people I met, about values and life styles that are not shared throughout this society. I came to realize that there are common concerns for many people that are alien to me. Several younger workers gave room and board money to their parents, and had been doing so ever since they had held part-time jobs in high school. I never gave my parents any room and board money, nor can I recall any of my childhood friends who did.

The people in the plant were much more closely tied to the local geographic area and to physical proximity with their families than my friends and I are. It was not uncommon to meet families like the Collinses, who have two generations working in the same plant. Most of the people at work had a number of relatives in the area. It was largely assumed that jobs would be taken in the immediate vicinity. This was not the case for the people I grew up with. Our sense was to find a job we liked wherever it would take us, and most often it took us away from where we had grown up.

There are many popular misconceptions and myths about blue collar work and so called blue collar lives. There is no need to caricature or romanticize these lives. There is much to admire in the real lives of these people and a great deal to learn from their capacities to make work more than a job.

EQUIPMENT
SHOPS
RE-CYCLE